D0971699

PROBLEMS OF NONSTOICHIOMETRY

Problems of
Nonstoichiometry

Editor:

A. RABENAU

Contributors:

J. S. ANDERSON

J. BENARD

E. F. BERTAUT

G. BERTHET

N. N. GREENWOOD

H. JAGODZINSKI

Y. JEANNIN

J. C. JOUBERT

P. J. L. REIJNEN

1970

NORTH-HOLLAND PUBLISHING COMPANY – AMSTERDAM · LONDON

Library of Congress Catalog Card Number 79–113752
North-Holland I.S.B.N. 0 7204 0173 9
American Elsevier I.S.B.N. 0 444 10047 4

PUBLISHERS:

NORTH-HOLLAND PUBLISHING COMPANY – AMSTERDAM
NORTH-HOLLAND PUBLISHING COMPANY, LTD. – LONDON

SOLE DISTRIBUTORS FOR THE U.S.A. AND CANADA:

AMERICAN ELSEVIER PUBLISHING COMPANY, INC.
52 VANDERBILT AVENUE
NEW YORK, N.Y. 10017

PRINTED IN THE NETHERLANDS

PREFACE

Nonstoichiometry had long been considered more a curiosity than a real problem of science until KURNAKOW, in the beginning of this century, recognized the practical importance of the ideas developed by BERTHOLLET more than a hundred years ago.

Further advances were made in the thirties with the work of SCHOTTKY and WAGNER who successfully put the problem on a quantitative basis for the first time.

Nowadays nonstoichiometry plays an important role in all branches of solid state research and physicists, chemists, crystallographers, ceramicists, metallurgists, etc. are actively involved with this subject.

The aim of this book is to disclose the problems and to make evident the trends. Experts, in treating selected topics on the subject from their point of view, reflect the state of affairs in the different fields of science.

Every discussion about nonstoichiometry involves thermodynamics. In the first chapter, thermodynamic and statistical mechanical considerations are discussed and a comparison is made of alternative attempts to provide general understanding of the subject. Special attention is given to the problem of grossly nonstoichiometric phases.

A comprehensive survey of recent work in the field is given in the next chapter which permits an analysis to be made of the new advances and ideas.

Besides thermodynamic considerations, there are the crystallographic aspects which are inseparably connected with nonstoichiometry. Spinels have been chosen as an example because this crystal system has been extensively investigated and the results are considered reliable and general.

The important role which methods of preparation play is demonstrated by the synthesis of a new group of metastable nonstoichiometric compounds using a topotactical exchange reaction.

Two-dimensional nonstoichiometric compounds may be formed by adsorption of a gas on a solid surface as applied in studies of metal–sulphur sysems. Of practical importance is the role nonstoichiometry plays in the sintering process of ionic solids. The theoretical model developed is in good agreement with experiments and offers new insight into the sintering process.

For studying complex problems of nonstoichiometry more use must be made in the future of new physical methods. The applicability of Mössbauer spectroscopy for this purpose is illustrated in the last chapter.

March 1970   ALBRECHT RABENAU
              Philips Forschungslaboratorium Aachen GmbH

CONTENTS

Contents

Contents

CHAPTER 1

THE THERMODYNAMICS AND THEORY OF NONSTOICHIOMETRIC COMPOUNDS

J. S. ANDERSON

Inorganic Chemistry Laboratory, Oxford University, Great Britain

1. Introduction

Perfect crystalline order automatically implies constancy of composition. Conversely, the fact that many binary and ternary inorganic compounds are nonstoichiometric shows that their crystal structures are aperiodically perturbed in some way, yet without losing their integrity. With the development of the statistical mechanics of crystals by Schottky and Wagner, Fowler and others, the notion of the perfectly ordered crystalline state gave place to that of statistical order, with an inbuilt measure of disorder, as the condition of inner equilibrium. What may now be termed classical point defect theory seemed to provide an explanation of nonstoichiometric compounds and does, indeed quantitatively, account for the properties of "well-behaved" substances, such as the typical semiconductor compounds, which are nearly, but not exactly, constant in composition. During the past fifteen years, however, a considerable weight of evidence has built up to show that it is impossible to interpret the chemically interesting, grossly nonstoichiometric compounds in terms of point defects – vacancies and interstitial atoms. To provide a satisfactory theory for these, two problems must be confronted: the real structure of grossly defective crystals and the thermodynamic criteria that such defective structures should represent the stable configurations.

Much of the evidence for the structure of nonstoichiometric compounds is dealt with in other chapters of this book. In this chapter, only such topics are referred to as are needed for an examination of the general theory. Limiting the discussion to binary compounds (though it can readily be

References p. 74

extended to ternary systems, such as the bronzes), those thermodynamic characteristics of nonstoichiometric phases will first be considered which any valid statistical mechanical treatment must reproduce. To be valid, a theoretical model must, equally, be compatible with what is known of structure. Depending upon the weight that is attached to the tendency to establish the maximum degree of order in crystals, and on the structural interpretation that is put on deviations from perfect order, theoretical models can be developed in several ways. We therefore compare the bases of alternative attempts to provide a general understanding for what is a large part of solid state inorganic chemistry.

2. Conditions of stability of nonstoichiometric compounds

2.1.

Nonstoichiometric compounds are essentially high temperature species. At low temperature it is not unusual to find that equilibrium is established between phases with a narrow, often infinitesimal, range of composition. Indeed, general considerations would suggest that true inner equilibrium at $0°K$ would require the disappearance of all residual, configurational entropy – i.e. the existence of fully ordered phases. (For examples see the Pr–O system, Fig. 1.1 and the U–O system, Fig. 1.2). At elevated temperatures, existence ranges usually become broader, even though for most compounds and especially for the ionic compounds of $[ns^2 np^6]$ cations, variations of composition may still be below the practical limits of analytical detection. In such cases they are revealed by the changes in electronic properties when a crystal is brought into equilibrium with varying partial pressures of its components. The refractory compounds, such as the oxides, are particularly important in this respect since, for these, the entropic contribution to the free energy can become large before crystalline order is lost at the melting point. Thus even thoria vaporizes incongruently at high temperatures; its composition range extends at least to $ThO_{1.998}$. In the present context, nonstoichiometric compounds are taken as those that have an analytically significant range of compositions. The distinction is one of convenience; the same compound may fall into both categories, at low and high temperatures respectively, and it must be inferred that a good theoretical model should provide a smooth transition from the state described by point defect theory to the more sophisticated description of grossly defective

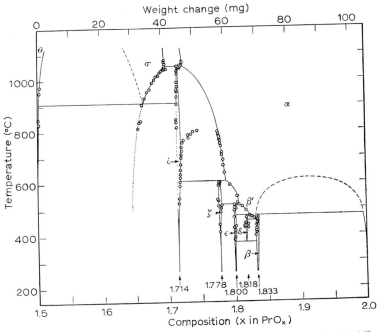

Fig. 1.1. The praseodymium–oxygen system (Hyde, Bevan and Eyring [1966]).

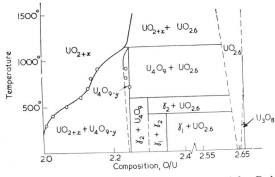

Fig. 1.2. The uranium–oxygen system $UO_{2.0}$–$UO_{2.7}$ (after Roberts).

structure. A second characteristic of nonstoichiometric compounds is that the chemical potentials of the components vary relatively gradually across the phase range: equilibrium with the components is bivariant, in the sense of

References p. 74

the phase rule. In an operational sense the existence of a nonstoichiometric phase must be established by two criteria: bivariance under equilibrium conditions and monophasic properties as established by X-ray difffraction, metallography etc.

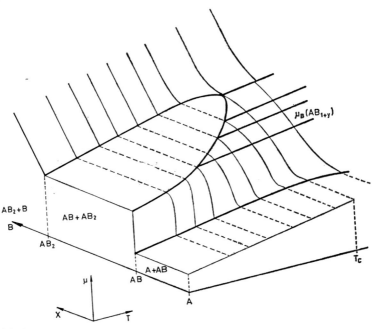

Fig. 1.3. 3-dimensional μ–T–X figure for a system with a nonstoichiometric phase AB_{1+y} and solvus curve closing at a critical temperature T_c. The $(\mu$–$T)_x$ lines are rectilinear on the curved free energy surface.

    As a prototype for general discussion, one may consider the equilibrium pressure–temperature–composition relations of a binary system A–B, in which there is a succession of phases: these are of "constant" composition at low temperatures and become nonstoichiometric at high temperature. If the three-dimensional p–T–X figure (or the equivalent μ–T–X figure) is projected on the T–X plane, the discontinuities in the p–X isotherms demarcate an envelope, within which the systems are univariant (Fig. 1.3); in many systems this envelope closes at some critical temperature T_c, above which a single nonstoichiometric phase exists, showing bivariant behaviour over a very wide composition range that spans the discrete phases stable at low temperature.

2.2.

It is convenient to discuss the stability and accessible composition range of nonstoichiometric compounds in terms of the free energy-composition curve. Figure 1.4 schematically represents the G–x relations (G = free energy

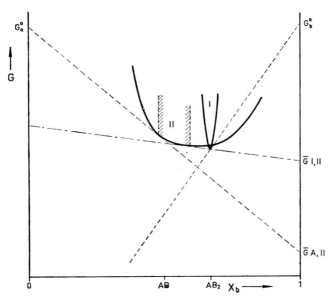

Fig. 1.4. Integral free energy curve for a nonstoichiometric phase coexisting with a phase of essentially invariant composition.

per atom in the phase $A_x B_y$) in a system containing a compound I of "constant" composition (AB_2) and a nonstoichiometric compound II with the nominal composition AB and a wide existence range. Defining the composition by the mole fractions

$$x_A = \frac{x}{x+y}, \qquad x_B = 1 - x_A = \frac{y}{x+y},$$

and the free energy by $G = x_A \bar{G}_A + x_B \bar{G}_B$, it is familiar that the tangent to the G–x curve defines the chemical potentials $\bar{G}_A$, $\bar{G}_B$ for any composition by its intercepts at $x_A = 1$, $x_B = 1$. The limits of the stable existence range of phase II are defined by coexistence with either free component A or phase I, at the same chemical potential, through the common tangent condition. The

References p. 74

composition range of phase I is similarly determined by coexistence with phase II or with free component B (at an arbitrary chemical potential $\bar{G}_B$ (max.) $> \bar{G}_B$ (I, II). If phase I is a compound of sensibly constant composition, the point of tangency is evidently only minimally displaced by a large

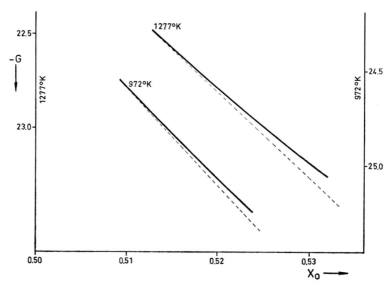

Fig. 1.5. Integral free energy curve for FeO (data of Ackermann and Sandford).

change in $\bar{G}_B$, and it follows that the free energy curve of the line phase is narrow and steeply dependent on composition. By contrast, $\bar{G}_B$ changes progressively across the composition range of phase II, and the $G–x$ curve for this phase must be relatively broad and flat. $G–x$ curves for the $Fe_{1-x}O$ and UO_{2+x} systems (Figs. 1.5 and 1.6), based on the measurements of Ackermann and Sandford [1966] and Roberts and Walter [1962] respectively, shows that this is indeed so.

  A satisfactory statistical mechanical model should quantitatively predict the form of the $G–x–T$ surface. The line phase I (Fig. 1.4) is presumably to be discussed by classical defect theory, with a low equilibrium concentration of defects; the high endothermicity of point defects and valence defects is clearly not so off-set by the configurational energy, TS_{config}, as to broaden the $(G–x)_T$ curve. The very different shape of the free energy curve for phase II implies either that defects are less endothermic or that the configurational entropy S_{config} is greater than calculated for a simple distribution of point defects over lattice sites. The free energy surface of phase I is a sharp blade in G–x–T space; that of phase II is a broad trough. It is the greater contribution from the entropy of the latter phase that results in a

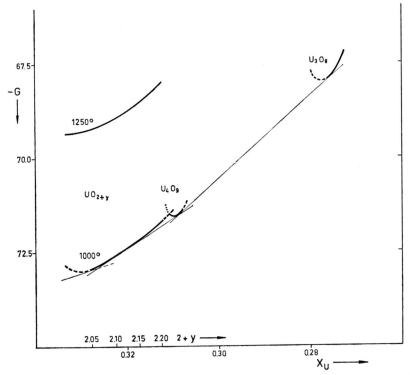

Fig. 1.6. Integral free energy curves for UO_{2+y}, U_4O_9 and U_3O_8 (data of Roberts and Walter).

mutual displacement of the two surfaces as the temperature is raised, and it is this relative displacement that broadens the existence range of II (Fig. 1.4).

For a simple regular solution system, the G–x curve would be symmetrical above the mid point, corresponding to the composition x_B^0, $1-x_B^0$ of the stoichiometric compound II. Brebrick [1967] and Kröger [1968] have pointed out that the G–x curves of wide-range nonstoichiometric compounds are asymmetric (Figs. 1.5, 1.6) and the asymmetry can be measured by the slope of the curve at the stoichiometric point: $(dG/dx)_0 = \mu(B)_0 - \mu(A)_0$.

References p. 74

2.3.

These chemical potentials depend on the energetics of the processes by which excess metal or excess nonmetal are incorporated. (i) There may be differences in the enthalpy or the entropy of the primary incorporation reaction arising from the ionization energies, energies of creating cation or anion vacancies, interstitials, etc., or entropy differences – e.g. in a Schottky solid

$$A\,(g) \rightleftharpoons A^x_M + V^x_B \qquad\qquad \text{incorporation of excess metal,}$$

$$\tfrac{1}{2}B_2(g) \rightleftharpoons B_g \rightleftharpoons B^x_B + V^x_A \qquad \text{incorporation of excess nonmetal.}$$

(ii) There may be differences in the free energies (hence in the equilibrium constants) for ionization of defects, resulting in different mean charge states

$$V^x_A \rightleftharpoons V'_A + e^+$$

$$V'_A \rightleftharpoons V''_A + e^+$$

$$V^x_B \rightleftharpoons V^{\boldsymbol{\cdot}}_B + e^- \qquad \text{etc.}$$

or (iii) differences in association or relaxation equilibria of the several kinds of point defect. The weight of evidence now available shows that this last factor may be particularly important.

In terms of the electronic structure of quasi-ionic solids, these considerations lead to an understanding of the differences between the compounds (e.g. oxides and chalcogenides) of the "normal" metals (i.e. with s^2p^6 or d^{10} cations) and those of the transition metals (with d^n cations, $n < 10$) and, more generally, the difference in behaviour of compounds derived from the highest and from the intermediate valence states of any element of variable valency (Fig. 1.7). In every case, to a tight-binding, localized-state approximation, an electron in the conduction band can be regarded as transforming a normal cation A^{z+} to a cation $A^{(z-1)+}$. The median energy of the conduction band can then be expressed as

$$E_c = I_z + B\,\frac{q^2}{r_{AB}} + \alpha$$

where I_z is the ionization energy $A^{(z-1)+} \to A^{z+} + e^-$, the second term is the change in Madelung energy and α a polarization energy; to this, for transition metal compounds, must be added a correction for change in the crystal field stabilization energy.

In compounds of "normal" elements, and of the highest valence state of transition elements, the uppermost filled band comes essentially from

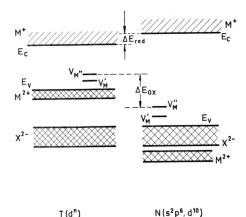

T(d^n)   N(s^{2}p^6, d^{10})

Fig. 1.7. Band diagram for oxides of transition metals and for s^{2}p^6 or d^{10} metals, controlling asymmetry of free energy curves.

occupied state of the anions – e.g. O^{2-} in oxides, X^- in halides, and positive holes reside mainly on these – e.g. as O^- or X^0. The median energy of the valence band is then given by $E_{v(N)} = A_{x-} - Bq^2/r_{AB} + \beta$, where A_{x-} is the electron affinity for $B^{(y-1)-} + e^- \rightarrow B^{y-}$ and β is again a polarization energy. In transition metal compounds of intermediate valence states it is generally accepted that the d^n states constitute a narrow band or set of discrete levels lying above the anion valence band, and the creation of positive holes then represents the ionization of A^{z+} (d^n) cations to $A^{(z+1)+}$ (d^{n-1}). This defines the position of the operative valence band by

$$E_{v(T)} = -I_{z+1} - \frac{Bq^2}{r_{AB}} + \beta' + \Delta \text{ (c.f.e.).}$$

The effective band gap is thereby considerably reduced, and the conversion of the stoichiometric compound to a nonmetal-rich compound, with the introduction of ionized metal vacancies and positive holes, is favourable in the case of transition metal compounds by the difference in free energy

References p. 74

ΔE_{ox} of the processes

$$B_{(g)} + B_B^x \rightarrow 2B_B^{\cdot} + V_A'' \qquad \Delta G_{ox,N}^0 \qquad (1.1)$$

$$B_{(g)} + 2\,A_A^x \rightarrow B_B^x + 2A_A^{\cdot} + V_A'' \quad \Delta G_{ox,T}^0 \qquad (1.2)$$

$$2\Delta E_{ox} = \Delta G_{ox,T}^0 - \Delta G_{ox,N}^0 .$$

In equations (1.1) and (1.2) $\mu(B)$ can be expressed in terms of the virtual chemical potentials of the structure elements of the crystal – normal or defect – appearing in the equations and, for the stoichiometric crystal, $2\,[B_B^{\cdot}] = [V_A'']$ or $2\,[A_A^{\cdot}] = [V_A'']$. It can then be readily shown that, at the stoichiometric point

$$\{\mu(A)_T - \mu(A)_N\}_0 = 2\Delta E_{ox}.$$

For the reduction process, or incorporation of excess metal, differences between the two classes of compounds arise chiefly from the crystal field stabilization energy term, and are small:

$$A(g) + A_A^x \rightarrow 2A_A^- + V_B^{\cdot\cdot}; \quad \Delta G_{red}^0 \text{ for both types } \{\mu(A)_T - \mu(A)_N\}_0 = 2\,\Delta E_{red}$$

Since the existence range of "normal" compounds is very small, there is probably little error in assuming that their G–x curves are roughly symmetrical about the stoichiometric point: $\mu(A)_{N,0} \approx \mu(B)_{N,0}$. Then, for transition metal compounds

$$\{\mu(B)_T - \mu(A)_T\} = (dG/dx)_0 \approx 2\{\Delta E_{ox} - \Delta E_{red}\} .$$

Insertion of the relevant quantities shows that this is, in all cases, strongly negative and indicative of a strong tendency for formation of nonmetal-excess, rather than metal-excess phases. For compounds of transition metals in high valence states, however, $\Delta E_{ox(T,N)}$ is small, whereas ΔE_{red}, though not large, is significant. $\{\mu(B)_T - \mu(A)_T\}_0 = (dG/dx)_0$ is then positive, and the form of the free energy curves dictates a preferential tendency for a range of existence on the metal-excess side of the ideal composition. Incorporation of excess nonmetal by creation of holes in the anion valence band is not excluded, and does indeed occur (e.g. at high Br_2 activity in the KBr–Br_2 system; see Kröger [1964]) but it is usually energetically too unfavourable to give rise to analytically significant nonstoichiometry, and there is no well established instance of such oxygen-excess phases in transition metal oxide chemistry.

It may be noted that where excess non-metal is incorporated in this way, relaxation displaces the X^0 ($=X_X^{\cdot}$) and an adjacent X^- ($=X_X^x$) ion towards each other, so that the positive hole is localized on what is essentially a X_2^- ion. Although there are no examples in transition metal chemistry, this is presumably the basis of the incorporation of an excess of oxygen, up to $BaO_{1.12}$ at 798°C, in barium oxide, as the equilibrium precursor of BaO_2 (Kedrovskii, Kovtunenko et al. [1967]).

2.4.

Stability limits have been considered thus far in terms of coexistence with adjacent phases. Outside the composition range thus set, the nonstoichiometric phase is unstable with respect to a decomposition reaction, into the coexisting pair, but is not absolutely unstable. If no crystal of the second phase is present, and if spontaneous nucleation is a kinetically hindered process, the nonstoichiometric phase may be formed – e.g. by a progressive

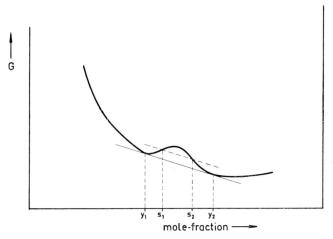

Fig. 1.8. Free energy curve for phase showing spinodal unmixing.

oxidation or reduction reaction – and may persist indefinitely with a composition outside the stable limit. It is metastable with respect to disproportionation, but transformation depends on the introduction of at least one viable nucleus of the product. There is an obvious risk that high temperature

References p. 74

nonstoichiometric states may be retained at lower temperatures, as metastable states.

The question then arises whether phase separation may in some circumstances be due to the absolute instability of a nonstoichiometric compound, rather than being contingent on control of the chemical potential by a second phase. It appears intuitively probable that there is some absolute limit to the tolerance of any crystal structure for defects. Spontaneous unmixing at low temperature and a broad range at high temperature would, indeed, give rise to just the rather symmetrical 2-phase envelope and critical miscibility temperature referred to above. The requisite conditions are shown in Fig. 1.8. At T_1 ($> T_c$) the G–x curve is of the form shown previously, over a very wide range of mole-fractions. The characteristic of this curve is that $dG/dx_B > 0$, $d^2G/dx_B^2 > 0$ everywhere. At a lower temperature T_2 ($< T_c$), the G–x curve develops two inflexions (the *spinodal* point) $d^2G/dx_B^2 = 0$, between which is a segment for which $d^2G/dx_B^2 < 0$. It can readily be shown that any composition between the spinodals is absolutely unstable, since a spontaneous fluctuation in composition leads to a decrease in free energy, and is therefore irreversible and selfpropagating (Cahn [1961]). The ultimate products of disproportionation are defined by the common tangent to the two segments of the curve, but the essence of spinodal decomposition is that it proceeds by small fluctuations of order and composition over large volumes, and is a continuous transformation, whereas nucleation involves a drastic fluctuation within a small volume, and maintains a discontinuity of structure and composition at the growth interface. It follows that the coexisting low temperature phase must have structures related topologically to each other and to the phase stable above T_c; the G–x curves are sections of a single free energy surface in G–x–T space and all thermodynamic functions must preserve their continuity in that surface. In nonstoichiometric systems with a critical temperature and intermediate phases which are structurally related – e.g. by variations in ordering pattern – there is a prima facie case for suspecting that the G–x–T surface will show spinodal instability.

It does not follow that the products of spinodal unmixing should both be stable phases in the binary system; one or both could be metastable with respect to some structurally unrelated phase. The iron–oxygen system probably exemplifies this case, and a highly schematic free energy diagram may well be as shown in Fig. 1.9. Above 570°, the FeO phase is stable over the range defined by its very flat G–x curve (Fig. 1.5) and coexists with iron or Fe_3O_4; the chemical potential of iron in $FeO_{1.0000}$ would apparently be greater than $\mu°(Fe)$. Below 570°, Fe and Fe_3O_4 are the stable phases but metastable wuestite can readily be obtained by quenching; its free energy curve lies everywhere above the $Fe–Fe_3O_4$ coexistence line. Manenc (Manenc, Vagnard and Benard [1962]; Herai et al. [1964]) has shown that, between 150° and 350°,

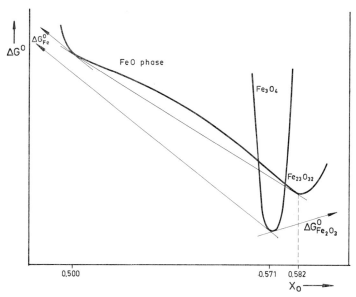

Fig. 1.9. Conjectural free energy curve for the Fe–O system at 200–350°C, showing spinodal unmixing into the $(2 \times 2 \times 2)$ cluster superstructure as metastable product.

quenched wustite undergoes unmixing into iron-rich and oxygen-rich regions, and that this is not a nucleation and growth process but proceeds cumulatively over extended regions. It gives rise, within a single crystal, to a regular periodic distribution of iron-rich and oxygen-rich lamellae, parallel to $\{100\}$ of the original crystal, and beautifully exemplifies Cahn's theory of spinodal unmixing, as producing fluctuations in composition with a regular and sinusoidal distribution in space. It may be surmised that the free energy curve at 350° is of the spinodal type shown, and that the oxygen-rich unmixing product is not the spinel Fe_3O_4, which requires nucleation, but the unknown metastable oxide based structurally on the ordering of the defect clusters in wustite. The simplest $(2 \times 2 \times 2)$ superstructure on this basis would have the composition $Fe_{23}O_{32}$. The iron-rich product of spinodal unmixing

References p. 74

could well be stoichiometric FeO which is also metastable with respect to the ultimate products.

3. Ordered phases versus nonstoichiometric compounds

3.1.

It is generally taken for granted that the experimental evidence proves the existence of compounds with widely variable composition, amongst ionic or quasi-metallic compounds as well as intermetallic compounds, so that the basic problem is to interpret their structure and behaviour in terms of solid state theory. By contrast, the work of crystallographers – notably A. Magneli and his school, A. D. Wadsley and B. T. M. Willis – has questioned whether the evidence is valid, or at least whether the theoretical problem has been rightly identified. The crystallographic viewpoint and much relevant factual material have been set out by Wadsley [1964] and recur in other chapters of this book.

It is indeed necessary to reappraise much of the experimental evidence in view of the fact that, in an increasing number of oxide and chalcogenide systems, apparently nonstoichiometric phases have been resolved, by prolonged annealing (which permits sluggish ordering and crystallite growth to proceed) and by more detailed crystallographic work, into successions of intermediate phases with well defined compositions and some common structural principle. The concept of compounds of variable composition is thereby supplemented, rather than replaced, by that of mixed valence compounds with "irrational" formulae determined by topological factors.

The crystallographic argument therefore is that minimization of the free energy of a crystal sets a high premium on perfect ordering, especially where coulombic forces contribute an important part of the lattice energy. Consideration of the relation between the structure of intermediate phases and that of a hypothetical parent structure with point defects shows that order is attained in two ways: (i) by *assimilation* of vacancies or interstitial atoms as structure elements of the crystal, or (ii) by *elimination* of point defects as a result of a local rearrangement of coordination polyhedra. In either case, formation of an intermediate compound of definite composition requires that the new structural arrangement is established with absolute regularity in three dimensions.

3.2. Superstructure ordering

Assimilation of point defects is well exemplified by the transition metal chalcogenides based on the B8 (NiAs) structure, reviewed more fully in Chapter 2.2. Earlier work (e.g. that of Haraldsen, 1936–39, on the chromium sulphides) suggested that there could be an almost complete and continuous transition between the B8 and C6 structures, by the interpolation of inter-stitial metal atoms in the C6 structure or the creation of vacancies in the B8 structure. Jellinek [1957] showed that prolonged annealing converted the "nonstoichiometric" CrS_x ($1.0 < x < 1.5$) into biphasic mixtures made up of well defined intermediates – Cr_2S_3, Cr_3S_4, Cr_5S_6, CrS – and similar results have been obtained in numerous other systems. These intermediate phases arise from the finite number of ways in which vacancies can be regularly ordered in alternate cation sheets of the B8 structure. In so far as it is per-missible to treat these compounds as ionic, calculations of the electrostatic lattice energy (Bertaut [1953], Burch [1969]) show that there is a strong energetic advantage in the complete ordering of unoccupied B8 sites within alternate cation sheets.

In a strict sense, therefore, these are not defective compounds; a dis-tinction must be drawn between a point defect vacancy and an unoccupied (potential) site. The latter is as much a structure element of the crystal as is an occupied site, and the crystal should be regarded as a pseudoternary system. Such a system clearly presents the possibility of order–disorder transitions of the classical kind; the switch of an atom from a "proper" site to a "proper vacancy" is in one sense substitutional disorder, in another sense the creation of an interstitial-vacancy point defect pair. The conse-quences of such order–disorder processes are considered further below.

This kind of vacancy ordering is not restricted to derivatives of the B8/C6 structures series. It is undoubtedly the basis of the intermediate phases in the oxygen–deficient fluorite structure oxides of cerium and praseodymium (Bevan [1955], Hyde, Bevan and Eyring [1966], Brauer and Gingerich [1957a, b]) and the related ternary compounds in the systems $LOF-LO_{1.5}$ (L = lantha-nide element) (Bevan, Cameron et al. [1968]). The low-temperature modifi-cation of TiO (Watanabe, Castles et al. [1967]) and the phase $VO_{1.27}$ (Westman [1960]), derived from the high temperature, highly nonstoichio-metric VO (Westman and Nordmark [1960]) also have ordered vacancy structures. Similarly, the ordering of interstitials is the basis of the series of intermediate oxides between UO_2 and U_3O_8. It appears also to underly the remarkable series of "microphases" found by Elliott (Elliott and Lemons [1963], Roof and Elliott [1965]) in the intermetallic cerium–cadmium compound $CeCd_{4.5}$, which raise the problem of just how small a change in the atomic ratios of the components may suffice to bring about a transformation from one crystallographically and thermodynamically distinguishable structure to another.

The feature common to such structures is that every occupied site is referable to a lattice site of some parent structure, with a smaller unit cell and higher symmetry, subject to only minor adjustments of atomic positions. Thus, in the monoclinic TiO, the Ti and O atoms, and the unoccupied sites, are distributed and ordered over the sites of a NaCl-type structure, which constitutes a sub-cell; the high temperature nonstoichiometric TiO phase is described in terms of the same, but defective and statistical, B1 cell. The intermediate phases have a fixed stoichiometry in so far as they have perfect order, particularly long range order. Relaxation of this constraint necessarily makes variability of composition possible and, with a variable degree of occupancy of lattice sites, the statistical sub-cell remains as the only repeating unit of the crystal.

There is ample evidence (Grønvold, Haraldsen and Vihovde [1954], Bøhm, Grønvold et al. [1955], Grønvold and Jacobsen [1956], Grønvold and Westrum [1959]) that order–disorder transitions take place at high temperatures in a number of systems, and are associated with progressive changes from line phases, through intermediate phases of measurable existence range, to grossly nonstoichiometric phases (compare the iron chalcogenides or praseodynium oxide). In such compounds as the B8/C6 chalcogenides, order may be lost in several ways:

(a) By an order–disorder transition within the incompletely filled sheets, while retaining the alternation of completely and incompletely occupied cation sheets.

(b) By randomizing the stacking of completely and incompletely filled sheets without disturbing the configuration of the latter.

(c) By distributing unoccupied sites randomly over all cation sheets. Of these modes of disorder, (c) approximates to a randomized point defect model; with (b), the stoichiometric balance between complete and incomplete layers is lost, so that interpolation of additional sheets of either kind would not further change the long range order parameters; with (a), both long and short range order are reduced.

In spite of a considerable number of investigations, the thermodynamics of these systems is not entirely clear. There are two distinct components in the ordering process: the distribution of cations and vacant sites over the totality of cation sites, and the ordering of cations in different valence and spin states. The latter determines the cooperative magnetic properties as well as making a large contribution to the coulombic energy. Transitions attributable to structural, electronic and magnetic ordering are, accordingly, observed at elevated temperatures, as lambda-type transitions. Only the onset of structural disorder relaxes the constraint on constancy of composition, and there is at present a lack of equilibrium data to show how far this results in nonstoichiometry. Most of the data relate to materials with compositions close to the ideal ordered structures. For Fe_7S_8, Lotgering [1955] concluded that ordering of spins is energetically more important than ordering of vacancies (Neel temperature 330°, order–disorder temperature for vacancies 300°C), but that for compositions off the ideal (e.g. for $Fe_{0.90}S$, which may not lie within the stoichiometric range of the pyrrhotite phase – Grønvold and Haraldsen [1952], Clark [1965]) vacancies are disordered below 200° although diffusion and attainment of order can take place readily at 150°C. With changes in ordering go changes in the crystallographic unit cell. Okazaki [1959] found for Fe_7Se_8:

T	a	b	c				
290°	$2a'\sqrt{3}$	$2b'$	$4c'$	Triclinic,	$\alpha 89.8°$	$\beta 89.6°$	$\gamma 90°$
300–350	$2a'\sqrt{3}$	$2b'$	$3c'$	Hexagonal			
350–450	$2a'\sqrt{3}$	$2b'$	$2c'$	Hexagonal			
450	$a'\sqrt{3}$	b'	c'	Hexagonal B8 sub-cell			

where $a'\sqrt{3}$, b', c' are the (orthorhombic) cell dimensions of the equivalent B8 cell. In addition to a magnetic transition around 450°K, which does not show in changes of the crystallographic unit cell, Grønvold [1968] observed a large entropy increase, peaking at 638°K, which can be associated with Okazaki's second transition. The magnitude of the energy increment is best compatible with disordering within the incomplete cation sheets whilst preserving alternation of stacking, and calorimetric measurements up to 1000°K show no evidence for the complete randomization of the structure, implicit in the unit cell found by Okazaki above about 700°K.

3.3. Elimination of defects: shear rearrangement

Intermediate compounds of definite composition may be derived in a second way from a hypothetical defective nonstoichiometric phase, by the process that Wadsley has termed crystallographic shear. In essence, this involves a rearrangement of the linkage between coordination polyhedra so that the anion: cation ratio is reduced within an infinite two-dimensional

Table 1.1

Principal types of shear structures

	Parent structure	Structure of shear planes	Examples
Based on MoO_3	Double ribbons of $[MO_6]$ octahedra joined into sheets by apex-sharing	$[MO_6]$ octahedra of adjacent ribbons linked by sharing edges	$Mo_{18}O_{52}$
Based on ReO_3	$[MO_6]$ octahedra joined in infinite 3-dimensional net by sharing apices	$[MO_6]$ octahedra joined in groups of 4 or 6 by sharing edges	Mo_8O_{23}, Mo_9O_{26}, $(Mo,W)_nO_{3n-1}$, $W_{20}O_{58}$
Based on TiO_2	Ribbons of edge-sharing $[MO_6]$ octahedra cross-linked by sharing apices	Adjacent ribbons of $[MO_6]$ octahedra joined by sharing faces	Ti_nO_{2n-1}, V_nO_{2n-1}
Block structures	ReO_3 structure, as above	Columns or blocks of ReO_3 structure spliced by sharing $[MO_6]$ octahedron edges to form two orthogonal sets of shear planes	Nb oxides from $NbO_{2.417}$ ($Nb_{12}O_{29}$) to $NbO_{2.500}$ ($Nb_{28}O_{70}$), ternary Nb–Ti and Nb–W oxides, Nb oxide fluorides

slab of the structure. Such slabs form the bounding faced of layers of more or less unaltered parent structure, and the composition is determined by two factors: the width of the layers of parent structure, variable within any one homologous series, and the rearrangement of coordination polyhedra which determines the composition of the shear plane and is the same for all compounds of any one series. If the resulting crystal is constant in composition, it must be inferred that the spacing between shear planes is absolutely regular. Table 1.1 sets out the characteristics of the main shear structures recognized at present.

Formation of these structures can be understood in terms of the elimination of a complete sheet of oxygen atoms. This probably begins with the collapse of a vacancy disc (Anderson and Hyde [1967]). The surrounding dislocation loop climbs through the crystal, draining it of oxygen vacancies until a complete sheet is eliminated, while the structure on either side joins up with a mutual displacement or shear along an appropriate crystallographic direction. This partial reconstruction, a sequential rather than a cooperative process, forms the characteristic structure of the shear plane. The result is to eliminate vacancies or interstitials from the structure, except at some low concentration characteristic of native disorder in ionic compounds, and responsible for diffusion processes. That they are not negligible at this limit is shown by the facility with which MoO_3 and V_2O_5 (which can be regarded as a double shear structure) undergo oxygen isotope exchange, a process dependent on self diffusion of oxygen ions.

The double shear or *block structures* are particularly relevant in the context of this chapter because they give rise to a great multiplicity of phases, with closely spaced atomic ratios. As can be seen from a projection along the shear planes (Fig. 1.10 a,b,c), they are built up from rectangular blocks (infinite columns) of ReO_3 structure, joined on all faces to analogous blocks (which are displaced vertically) so as to form the interface out of a sheet of edge-shared octahedra. Each block is n octahedra wide, m octahedra long; the blocks may either be isolated or linked by edge sharing of octahedra at the same level in groups of p blocks (Fig. 1.10 b,c). In the known structures, $p = 1$, 2 or ∞. The composition of a block is given by the expression

$$M_{nmp}O_{3nmp - p(n+m)+4}$$

and the composition of the crystal, given by the sum of a pair of such expressions with n, m, p, n', m', p' for the two sets of blocks at levels 0 and $\frac{1}{2}$, can be represented by the symbol

$$(n \times m)_p + (n' \times m')_{p'}.$$

Subject only to two constraints – preservation of anion–cation charge balance and the geometrical requirement that the two sets of blocks should fit together in a space-filling pattern – every conceivable set of symbols p, m, p, n', m', p' represents a possible compound of unique structure. The structural principle thus admits of an enormous number of phases, with exceedingly closely spaced compositions and, by inference, correspondingly small differences in thermodynamic properties. Charge balance can be achieved

References p. 74

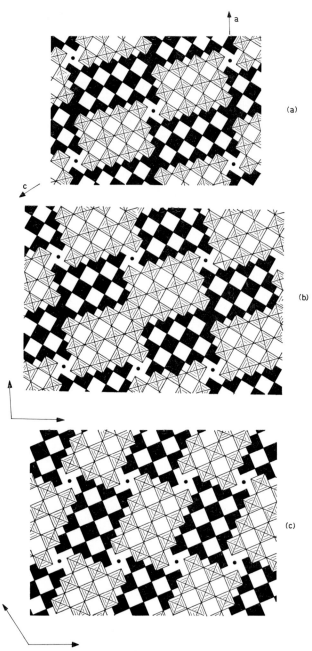

Fig. 1.10. Double shear or block structures. (a) $Nb_{28}O_{70} (3 \times 4)_1 + (3 \times 5)_\infty$; (b) $TiNb_{24}O_{62}$ or $Nb_{25}O_{62} (3 \times 4)_2 + (3 \times 4)_2$; (c) $WNb_{12}O_{35} (3 \times 4)_1 + (3 \times 4)_1$.

by substitution of Nb^{4+} for Nb^{5+} in mixed valence phases with the total
O:Nb ratio <2.5, in ternary phases by introduction of Ti^{4+} for Nb^{5+}
(O:M <2.5) or W^{6+} for Nb^{5+} (O:M >2.5), or by substitution of F^- for
O^{2-} (O+F:M >2.5). A few of the compounds that have been characterized
as distinct phases are listed in Table 1.2.

Table 1.2

Block structures in niobium oxides and ternary oxides

Compound	Structure symbol	Anion: cation ratio
Niobium oxides		
$Nb_{12}O_{29}$	$(3 \times 4)_\infty$	2.417
$Nb_{22}O_{54}$	$(3 \times 4)_1 + (3 \times 3)_\infty$	2.454
$Nb_{15}O_{37}$	$(3 \times 5)_\infty$	2.467
$Nb_{47}O_{116}*$	$\{(3 \times 4)_1 + (3 \times 3)_\infty\} + \{(3 \times 4)_2\}$	2.468
$Nb_{32}O_{79}$	$(3 \times 4)_1 + (3 \times 3)_2$	2.469
$Nb_{25}O_{62}$	$(3 \times 4)_2$	2.480
$Nb_{35}O_{87}$	$(3 \times 4)_2 + (3 \times 3)_1$	2.486
$Nb_{53}O_{132}*$	$\{(3 \times 4)_1 + (3 \times 5)_\infty\} + \{(3 \times 4)_2\}$	2.490
$Nb_{28}O_{70}$	$(3 \times 4)_1 + (3 \times 5)_\infty$	2.500 $H-Nb_2O_5$
$Nb_{16}O_{40}$	$(4 \times 4)_\infty$	2.500 $N-Nb_2O_5$
Ternary oxides		
$TiNb_2O_7$	$(3 \times 3)_\infty$	2.333
$Ti_2Nb_{10}O_{29}$	$(3 \times 4)_\infty$	2.417
$TiNb_{24}O_{62}$	$(3 \times 4)_2$	2.480
$WNb_{12}O_{33}$	$(3 \times 4)_1$	2.538
$W_4Nb_{26}O_{77}*$	$\{(3 \times 4)_1\} + \{(4 \times 4)_1\}$	2.563
$W_3Nb_{14}O_{44}$	$(4 \times 4)_1$	2.588
$W_5Nb_{16}O_{65}$	$(4 \times 5)_1$	2.619
$W_8Nb_{18}O_{69}$	$(5 \times 5)_1$	2.635
Oxide fluorides		
Nb_3O_7F	$(3 \times \infty)_\infty$	2.667
$Nb_{17}O_{42}F$	$(3 \times 5)_1 + (3 \times 6)_\infty$	2.530
$Nb_{31}O_{77}F$	$(3 \times 5)_2$	2.516

It would appear from the work of Andersson [1967] (compare also
Roth and Wadsley [1965], Wadsley [1967]) that any synthetic mixture with
the appropriate composition in these niobium oxide systems is capable of
reaction to produce a distinct block structure, and that intermediate compo-

sitions form biphasic mixtures of two block structures rather than nonstoichiometric compounds. It follows that the fully ordered structures are energetically advantageous, compared with disordered structures, even for small changes in composition. Indeed, the succession of phases may be made even closer by the formation of regular 1:1 intergrowth structures (marked * in Table 1.2) in place of biphasic mixtures or nonstoichiometric phases.

The thermodynamic implications of this synthetic and structural work are important for the present discussion. No diffusion measurements have yet been made on shear structure compounds, but the compounds are interconvertible by oxidation or reduction at high temperatures, and this process requires diffusive reconstruction, with the migration of shear planes to new regular spacings throughout the crystal. A single crystal of $H-Nb_2O_5$ retains its integrity during reduction, and traverses the stages

$$Nb_{28}O_{70} \rightarrow Nb_{35}O_{87} \rightarrow Nb_{25}O_{62} \rightarrow Nb_{32}O_{79} \rightarrow Nb_{22}O_{54} \rightarrow Nb_{12}O_{29}.$$

The extent to which these reactions involve reconstruction of the unit blocks can be seen from Table 1.2. Although some workers have considered that the shear phases have extended composition ranges, Khan (Anderson and Khan [1969]) has found no detectable composition ranges for the titanium and vanadium Magneli phases up to 1250°C; the niobium oxide block structures were synthesized at 1350–1400°C. Their behaviour thus appears to be that of typical line phases in the equilibrium diagram, even at high temperatures.

We can therefore postulate a $G-x$ diagram as shown in Fig. 1.11, with small, closely spaced but genuine steps in $\bar{G}_{O_2}$, which could readily be mistaken for a continuous bivariant curve if experimental measurements are too widely spaced and if adequate crystallographic data are lacking. In the integral free energy diagram, each phase has its own $G-x$ curve, and the evidence summarized above suggests that these curves will be rather narrow. The envelope of these curves, formed by the coexistence condition tangents, must inevitably approximate to the smooth, low curvature $G-x$ curve typical of a single phase of variable composition. The existence of ordered intergrowth phases demonstrates that there is a free energy gain from such inter-phase ordering (curves A, B and AB in Fig. 1.11), and we must enquire whether there is any thermodynamic limitation – as distinct from the kinetic difficulty of highly complex ordering processes – on the multiplicity of phases that might exist, and what meaning is to be attached to the concepts of a "defined compound" and a "randomized structure".

The free energy of a crystal is a statistical quantity, related to the partition function, being the value assigned to the most probable configuration of a large assembly of atoms. Fluctuations about this mean may occur and, in a closed system (a crystal of fixed total composition at constant temperature)

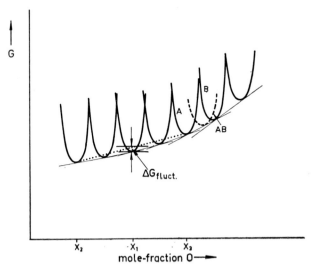

Fig. 1.11. Form of integral G–x curve for system with close succession of intermediate phases. AB is a regular intergrowth phase derived from the structures of phases A and B. The phase of composition x_1 disproportionates, by a fluctuation, into the compositions x_2 and x_3.

a configurational fluctuation may take the form of a change in composition *within a small region*, as a result of random atom movements into or out of some element of volume. The probability of such a fluctuation depends (a) on the temperature and the free energy change involved in accordance with a distribution function of the form $w = \exp[-\Delta G/kT]$, and (b) on the size of the region considered, since the probability of an outflow or inflow of ΔN atoms, to give a composition change $\Delta N/N$, is proportional to $N^{-1/2}$. In Fig. 1.11 the form of the free energy envelope for a succession of phases is such that there is only a small increase in free energy if a preparation with the composition x_1 undergoes a highly local unmixing into two new structures corresponding to the compositions x_2 and x_3. At high temperatures such fluctuations must become significant on a microscopic scale; their compositional amplitude and the size of the regions affected must both

References p. 74

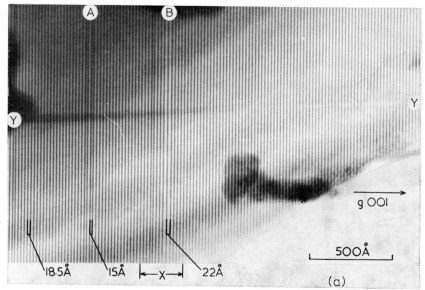

Fig. 1.12. Electron microscope observations on $W_3Nb_{14}O_{44}$ (Allpress, Sanders and Wadsley). (a) Fringe imaging of lattice planes, showing occasional interpolation of occasional shear plane spacings of 15 Å and 22 Å, in place of the regular 18.5 Å spacing of the long dimension of (5×4) blocks. (b) Microdensitometer trace and interpretation of a comparable faulted region in a crystal of $W_4Nb_{26}O_{77}$.

increase steeply with temperature. This consideration, depending on the thermodynamics of particular chemical systems, sets a limit on the number and complexity of the phases that can exist, and also on the precision with which a macroscopic crystal can define its composition and structural perfection.

Recent electron microscopy by Allpress, Sanders and Wadsley [1969] bears directly on this point. The cell dimensions of the block structures are large enough to permit the principal planes to be observed directly, by fringe imaging, whereby the spacing of the shear planes can be measured directly, from point to point within a single crystal. The scale of structure identification is thereby changed, from the statistical level of the single crystal, practically to the level of the individual unit cell. Observations on the niobium tungsten oxides – using the actual specimens employed in the crystal structure determinations – showed that whereas the fringe spacing throughout the bulk of the crystals accorded exactly with the shear plane separations inferred from the crystal structure, the crystals showed occasional "faulted"

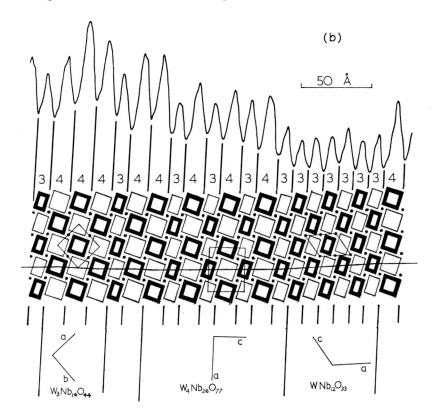

regions in which the fringe spacing changed abruptly from the "proper"
value to another, indicative of and identifiable as a local change in block size.
Thus, within a region of a $W_5Nb_{16}O_{55}$ crystal measuring about 2500×1500
Å the regular fringe spacing of 18.5 Å (corresponding to the long dimension
of the 5×4 blocks) was interrupted by four spacings of 15 Å, characteristic
of the 4×4 blocks of $W_3Nb_{14}O_{44}$, and two spacings of 22 Å, corresponding
to the 6×4 blocks of the (so far unknown) $W_7Nb_{18}O_{66}$ structure (Fig. 1.12
a,b). Over a region about 130 unit cells wide, shown in the micrograph, the
stoichiometry appears as

$$125(W_5Nb_{16}O_{55}) + 4(W_3Nb_{14}O_{44}) + 2(W_7Nb_{18}O_{66})$$

or $MO_{2.6187}$, as compared with the formula $MO_{2.6190}$ for fault-free
$W_5Nb_{16}O_{55}$. The regions of anomalous composition can thus be regarded

as true fluctuations within a crystal of fixed, nearly ideal stoichiometry. They may be relatively extensive in at least one dimension, or very small.

It is noteworthy that such frozen fluctuations were found less frequently in the binary niobium oxides than in the ternary oxides. This suggests that fluctuations in the distribution of alter-valent cations determine the local structure; in binary mixed valence compounds these can be redistributed non-diffusively, by electron transfer. The distinction is important, and indicates dangers in extrapolating ideas from the properties of ternary or "doped" binary compounds; it draws attention also to the possible role of quite low levels of impurity atoms in the behaviour of nonstoichiometric systems.

Some caution must be exercised, however, in drawing inferences about the structures at high temperatures from the crystallographic information. The degree of randomness, the size of and amplitude of fluctuations and the role of classical point defects must all increase with temperature. Some degree of ordering, which may result in nucleation and segregation of regions with slightly aberrant structure and composition, inevitably takes place during the most rapid cooling, before diffusion processes are frozen, and the size of the regions which may be changed in composition and denuded of point defects or extended defects will depend on the mean diffusion length in a particular experiment. Thermodynamic equilibria of nonstoichiometric phases are measured (and measurable) only at temperatures where diffusion processes enable non-equilibrium states to relax. Under these conditions there may be little operational meaning in the question "what is the true structure of a defect phase?": the answer must depend both on the size of the sample analysed (as considered above) and on the time scale inherent in the method of investigation. The occurrence of diffusion implies that any lattice site is occupied part of the time and vacant part of the time; the site occupancy fractions, which may in favourable cases be measured as part of crystal structure determinations, represent the – randomly distributed – fraction of sites in which an atom is found at room temperature, when diffusion is more or less immobilized. Under high temperature steady state conditions it represents the fraction of the time for which a given site is occupied. Atoji and Kikuchi [1968] have calculated from the diffusion parameters that, in $ZrC_{0.98}$, described as having a truly random defect structure, the mean lifetime of a carbon atom on a given octahedral site decreases from 2.2×10^7 sec at 700°C to 22 sec at 1000°C, 4.2×10^{-3} sec at 1300°C and 2.5×10^{-6} sec at 1700°C. It is only when this mean lifetime is long enough that questions about order, even at the unit cell or nearest neighbour level, are meaningful;

when the mean lifetime is short, structure and short range order could be interpreted only in terms of some correlation between the occupancy of groups of sites. At present we have few means of investigating this.

4. Nonstoichiometry as an order–disorder problem

4.1.

The preceding section makes it evident that ordered configurations are energetically advantageous, especially in ionic crystals where the coulombic interactions are long range forces. Nonstoichiometry is associated with a breakdown of perfect order. The basic question is therefore the extent to which order – particularly short range order – is retained in a crystalline solid which is, in every operational sense, monophasic and grossly defective.

States of order in a crystal are conveniently specified by a set of ordering parameters, which may be defined in several ways. The most general form is a pair correlation coefficient $P_{mn}(r_{ij})$, measuring the probability that site j, at a distance given by vector r_{ij} from an atom m at site i, is occupied by a particular kind of structure element n (an atom or a vacancy). If the perfect crystal at $0°K$ is taken as the reference state, and is divided into an appropriate set of sublattices (Wojciechowski [1961]), $P_{mn}(r_{ij})$ has some value between 1, for perfect order (all structure elements on proper sites) and x_n for a completely random distribution of species n, which makes up the mole-fraction x_n of the crystal. Thus, for a Frenkel type nonstoichiometric crystal AB_{1+y}, with mole-fractions x_v of cation vacancies, x_i of interstitials, the A sublattice would be properly occupied by A atoms, the interstitial sublattice entirely by vacancies; if the point defects are randomly distributed, $P_{AA}(r) = 1 - x_v$, $P_{AI}(r) = x_i$. This random distribution is the basis of classical defect theory.

A distinction must be drawn, however, between long range order and short range order. $P(r_{ij})$ may vary with r_{ij}, and may fall off to the random value when r_{ij} becomes equal to several or many lattice spacings, even though there is a high degree of perfection in the nearest neighbour environment of each site. In this case, $P_{mn}(r)(\text{SRO}) > P_{mn}(r)(\text{LRO})$, but as long as the correlation is taken between sites that are not near neighbours the distribution of defects approximates to randomness. There is another possibility: that the crystal retains very strong elements of local order, and is composed of regions, of unspecified size, throughout which ordering is nearly perfect, but which abut on analogous regions with which they are uncorrelated. In inter-

References p. 74

metallic compounds the ordering pattern in adjacent regions may be of the same kind; they are separated by an antiphase boundary. In nonstoichiometric compounds, adjacent regions (*microdomains*) would need to differ in composition, as well as in the consequential pattern of ordering. In hypothetical structures of this kind, long range correlation would be completely random, but short range disorder would be restricted to sites on the microdomain boundaries.

Although order–disorder phenomena in intermetallic systems have been extensively studied (cf. Elcock [1956]) and such related problems as the cation distribution in spinels and other ternary compounds have been dealt with (cf. Borghese [1967]), relatively little attention has been given to non-stoichiometric compounds. These invariably present a multiple sublattice problem of some complexity, which can be simplified to some extent because disorder and variable site occupancy are restricted to a certain set of sublattices within a constant matrix. Thus, in the B8/C6 transition phases, the nonmetal sublattice is unaffected to a first approximation; as long as the sequence of filled (f) and defective (d) sheets follows the order ..f d f d... characteristic of the hexagonal end members of the series, the filled sheets can also be neglected. The problem thus reduces to the two-dimensional ordering of a variable population of filled sites and empty sites in the defective sheets. Even so, the compositions and structures of the main intermediate phases in this series represent a 12-sublattice problem. Similar analyses of the minimum requirements for specifying an order–disorder treatment can be made for other structures, and could be so framed as to include defect clusters and polyatomic groups as structure elements.

In principle, therefore, it should be possible to formulate the grand partition function for a crystal (or for a specific set of sublattices of a crystal) with a variable composition, including the pair correlation coefficients of long range order and a nearest neighbour site preference energy to control short range order. However, the ordering problem has been conveniently solved to this degree of approximation only for simple, fixed stoichiometries that simplify the cumbersome equations. Multiple sublattice systems of variable composition (e.g. ordering in the Au–Cu alloys) have been treated in a lower approximation including long range order only, but short range order is unquestionably so important in ionic compounds that there is little profit in pursuing a model that ignores it. It is therefore at present impracticable to relate the thermodynamics of nonstoichiometric compounds to an order–disorder model.

4.2.

The relation between different ways of framing a theoretical model can be summarized as in Fig. 1.13. At one extreme is the high degree of short range and long range order that replaces nonstoichiometric phases by closely

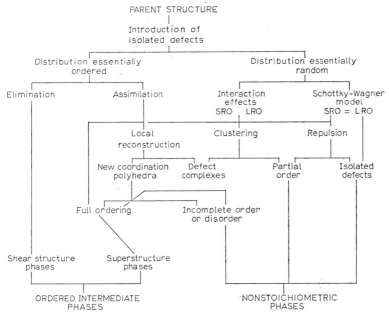

Fig. 1.13. Chart showing relationship between random point defects, interactions and ordering.

spaced series of definite compounds, in which defects are negligible. At the other extreme is the random defect model. Between these extremes, interactions between defects operate to reduce the degree of randomness and can result in the configuration of minimum free energy being highly ordered. Statistical mechanics could either start from a random defect model, introducing a configurational free energy and finding a tendency towards order for certain compositions and temperatures, or from a fully ordered crystal as the reference state, introducing disorder as a perturbation. No satisfactory model of the latter kind has yet been framed, for the reasons discussed above, but it is worth examining one form of it: the concept of perfect short range

References p. 74

order of two types, in uncorrelated microdomains. This has been qualitatively accepted in recent years by a number of workers, but its thermodynamic
implications have not hitherto been critically considered. Section 8.4 attempts
to formulate the hypothesis more explicitly than hitherto.

It should be noted that most of the statistical mechanical treatment of
defects is not invalidated if point defects are replaced by extended defects
or defect clusters. Dilute defects in uni-univalent ionic crystals – e.g. cation
vacancies in KCl – may retain the point group symmetry of the lattice site
with which they are associated. More generally, even for such systems but
especially where defects bear a higher charge and are associated with altervalent ions, relaxation processes may impose a local structural reorgani-
zation and, possibly, an association of defects into multiplets (cf. Lidiard
[1962]). This is the essence of the defect clusters in UO_{2+x} or $Fe_{1-x}O$.
Creation of nominal point defects is an endothermic process, and the energy
of defect creation, as determined operationally, includes the energies of
relaxation. The configurational entropy of clusters may well introduce new
features – e.g. an orientational entropy – but, provided that defect clusters
can be treated as units to be randomized in the crystal, the model based on
point defects must be changed only in detail, not in substance.

5. Random point defects

5.1.

The simplest approach to nonstoichiometry is in terms of an unbalance
between the native point defects of a crystal. In principle, a crystal AB_n may
embody every mode of disorder: vacancies on A and B sites, A and B atoms
in interstitial positions, substitution of A atoms in the B sublattice and vice
versa. Each such perturbation of perfect order increases the internal energy
of the crystal, regarded as a closed system, by an increment E_{ij}, where i
indicates the type of point defect (vacancy V, interstitial I, substitutional S)
and $j = A$ or B. To a first approximation, these defect energies are taken as
having constant values, independent of the temperature and total concentration of defects; they are not readily or reliably calculated from first
principles, except perhaps for such nearly ideal dilute defect systems as the
alkali halides, and the values assigned to them are best regarded as empirical
parameters, derived from the combination of suitable data – e.g. from the
temperature coefficient of ionic conduction in stoichiometric and doped

Table 1.3

Crystal	Point defect pair	Enthalpy of formation of conjugate point defect pairs ΔH e.v. (observed)	ΔH e.v. (calc.)
AgBr	$V'_{Ag} + V^{\cdot}_{Br}$	1.5–2.2	—
	$V'_{Ag} + Ag^{\cdot}_{I}$	0.87–1.27	—
AgCl	$V'_{Ag} + Ag^{\cdot}_{I}$	1.08–1.60	—
KCl	$V'_{K} + V^{\cdot}_{Cl}$	2.1–2.3	2.2
KBr	$V'_{K} + V^{\cdot}_{Br}$	1.96–2.0	1.92
NaCl	$V'_{Na} + V^{\cdot}_{Cl}$	2.02–2.19	1.91–2.12
CaF$_2$	$V^{\cdot}_{F} + F'_{I}$	2.8	2.7
MgO	$V''_{Mg} + V^{\cdot\cdot}_{O}$	—	4–6
CaO	$V''_{Ca} + V^{\cdot\cdot}_{O}$	—	3–5
Al$_2$O$_3$	$2V_{Al} + 3V_{O}$	20.4	—
UO$_2$	$V^{\cdot}_{O} + O'_{I}$	3.44	—
PbS	$V^{x}_{Pb} + V^{x}_{S}$	2.4	
	$V'_{Pb} + V^{\cdot}_{S}$	1.75	

crystals. Some theoretical and experimentally determined energies for creation of conjugate defect pairs in simple binary crystals are given in Table 1.3. Disorder in a closed system can be represented as the consequence of quasi-chemical disordering reactions:

$$0 \rightleftharpoons V_A + V_B \qquad E_1 = E_{VA} + E_{VB} \qquad (1.3)$$

$$A_A \rightleftharpoons V_A + A_I \qquad E_2 = E_{VA} + E_{IA} \qquad (1.4)$$

$$A_A + B_B \rightleftharpoons A_B + B_A \qquad E_3 = E_{SA} + E_{SB} \qquad (1.5)$$

etc.

In an open system – a crystal in equilibrium with the components A or B at a specified chemical potential – some transfer or nonstoichiometric incorporation reaction is possible

$$A(g) \rightleftharpoons A_A + nV_B \qquad E_4 = E_{XA} + nE_{VB} \tag{1.6}$$

$$A(g) \rightleftharpoons A_I \qquad E_5 = E_{XA} + E_{IA} \tag{1.7}$$

$$\tfrac{1}{2}nB_2(g) \rightleftharpoons nB_B + V_A \qquad E_6 = nE_{XB} + E_{VA} \tag{1.8}$$

$$A_I + \tfrac{1}{2}nB_2(g) \rightleftharpoons A_A + nB_B \qquad E_7 = nE_{XB} - E_{IA} \tag{1.9}$$

etc.

and the real situation corresponds to every combination of such internal and transfer equilibria, which are thermodynamically inter-dependent. In (1.8) and (1.9), the nonmetal component is represented as present in the gas as diatomic molecules, as is appropriate for the halogens, oxygen and sulphur at high temperatures. The change in internal energy in reactions (1.4) to (1.9) has two component parts: one due to the change in the number of point defects, the other involving the changes in binding energy of the crystal – i.e., an effective valence change. For ionic crystals, this can be related to changes in the ionization state of one or other component, as discussed earlier. For quasi-metallic or covalent crystals, it will depend on the population of electron band states and be measured by the shift of the Fermi level. It will be noted that processes such as (1.4.), (1.5) and (1.7) are conservative in lattice sites whereas the number of sites of each kind, B_A, B_B ($=nB_A$) is augmented by processes such as (1.3), (1.6), (1.8) or (1.9). The deviation of a crystal $AB_{n+\delta}$ from the ideal stoichiometry can then be expressed as the excess or deficiency of B atoms per unit volume:

$$\delta = [B_B] + [B_I] - [V_B] + [B_A] - [A_B] - n\{[A_A] + [A_I] - [V_A] + [A_B] - [B_A]\}. \tag{1.10}$$

The equilibrium state can then be derived from the grand partition function in which, for comparison with experimental data, it is usually convenient to take the chemical potential of the nonmetal as an environmental variable:

$$\Xi(N_A, \mu_B, T) =$$

$$\sum_{\Delta N_B} \left(\prod_{ij} \Omega_{ij} \exp[-N_{ij}E_{ij}/kT] \right) [\kappa(T) \cdot \lambda_B]^{\Delta N_B} \exp[-\Delta N_B E_{XB}/kT] \tag{1.11}$$

by the usual process of finding the largest term and setting $(\partial \log \Xi/\partial N_{ij}) = 0$.

In (1.11), ΔN_B is the stoichiometric excess of B atoms in a crystal with a constant number N_A atoms of A; N_{ij} the number of defects of each type ij, Ω_{ij} is the configurational degeneracy of the state with N_{ij} defects distributed over the lattice sites, $\kappa(T)$ is the contribution of each excess B atom to the vibrational partition function of the crystal and $\lambda_B = \exp(\mu_B/kT)$.

In the solution of this general, but cumbersome, equation there is inevitably a large disparity between the concentrations of different defect types, since these involve factors of the form $\exp(-E_{ij}/kT)$. It is on this basis that the usual approximation is made, that native disorder and non-stoichiometry can be referred to one conjugate defect pair and a single inner equilibrium – usually (1.3), (1.4) or the converse of (1.4) with interstitial B atoms. This approximation undoubtedly breaks down for many substances at temperatures approaching the melting point, as the defect concentration becomes large (e.g. 10^{-3}).

In terms of this approximation, all the essentials of the model can be illustrated by the example of a crystal of ideal formula AB, with Schottky disorder. The crystal contains $B = N_A + N_{VA} = N_B + N_{VB}$ sites of each kind, and the stoichiometric defect δ (in $AB_{1+\delta}$)

$$= (N_B - N_A)/N_A = (N_{VA} - N_{VB})/N_A.$$

$$N_{VA}/B \approx N_{VA}/N_A = p_{B_2}^{\frac{1}{4}} \kappa(T) \exp[-E_{XB}/kT] \tag{1.12}$$

$$N_{VB}/B \approx N_{VB}/N_A = \frac{1}{p_{B_2}^{\frac{1}{4}} \kappa(T)} \exp[-(E_{VA} + E_{VB} - E_{XB})/kT]. \tag{1.13}$$

For the stoichiometric crystal in equilibrium with B_2 vapour at the pressure $p_{B_2}(0)$

$$[V_A] = [V_B] = \xi = \exp[-(E_{VA} + E_{VB})/kT] = K_s^{\frac{1}{2}} \tag{1.14}$$

$$\text{whence } p_{B_2}^{\frac{1}{4}}(0) = \frac{\xi}{\kappa(T)} \exp[E_{XB}/kT]. \tag{1.15}$$

The change in composition when the crystal is equilibrated with a partial pressure $p_{B_2}(\delta)$ is then given by

$$\left(\frac{p(\delta)}{p(0)}\right)^{\frac{1}{4}} = \exp\left[\frac{\mu_B(\delta) - \mu_B(0)}{kT}\right] = \frac{\delta}{2\xi} + \left[\left(\frac{\delta}{2\xi}\right)^2 + 1\right]^{\frac{1}{2}}. \tag{1.16}$$

The deviation from stoichiometry produced by a given change in chemical

potential depends directly on the degree of intrinsic disorder in the stoichiometric crystal.

For $\delta \gg \xi$, stoichiometric deviation is effectively due to only one kind
of defect: nonmetal vacancies on the metal-rich side, metal vacancies on the
nonmetal-rich side. As noted earlier, these involve different creation processes and the corresponding partial molal enthalpies for the nonstoichiometric reaction differ on either side of the ideal composition.

For $\delta > 0$: nonmetal-rich $\qquad \bar{H}_{B_2} = E_{XB}/2k$
$\quad \delta < 0$: metal-rich $\qquad \bar{H}_{B_2} = (E_{XB} - E_{VA} - E_{VB})/2k$.

The Schottky Wagner treatment of point defects is essentially quasi-
chemical; it derives the equilibrium constants for the internal set of reactions
between structure elements and the displacement of that equilibrium by
reaction with the components. It is incomplete in that it takes no account of
electronic states within the crystal and (as Kröger has pointed out) the
precise meaning attached to a point defect depends on the assumption made
about the charge state of the atoms in the normal crystal. Thus, in developing
the theory, Wagner treated Cu_2O_{1+x} as an ionic crystal in which positive
holes were localized equally probably on any cation site, as Cu^{2+} ions:

$$\tfrac{1}{2}O_2 \rightleftharpoons O_O^{2-} + 2V_{Cu} + 2\,e^+$$

$$2Cu_{Cu}^+ + 2e^+ \rightarrow 2Cu^{2+}\,.$$

The nett reaction $2Cu^+ + \tfrac{1}{2}O_2 \rightleftharpoons 2Cu^{2+} + O^{2-} + 2V_{Cu}$, with $[V_{Cu}] = [Cu^{2+}]$
then leads to a stoichiometric excess of oxygen proportional to $p_{O_2}^{\frac{1}{8}}$. Similarly,
in hypostoichiometric ZnO_{1-y}, the pressure-dependence of composition will
differ according to the assumptions made about the localization of excess
electrons as Zn^+ or Zn^0.

An ion vacancy, as an interruption in the regular periodicity of charges
in the crystal, bears a virtual charge equal and opposite to the charge that is
proper to the site. An anion vacancy is thus a trap for electrons, a cation
vacancy for positive holes; the charges can be described either in terms of
hydrogen-like wave functions centred on the vacancy, or as reduced or
oxidized ions on nearest neighbour sites. Detachment of a trapped charge
from a point defect, to form a mobile electron or positive hole, requires an
increment of energy.

5.2.

In the Kröger-Vink symbolism, a consistent description, independent of the ionicity of the normal structure elements, is provided by regarding the displacement of neutral atoms as the basic process of creating point defects, so that every lattice site bears its normal charge, relative to the perfect crystal. Detachment of electrons or positive holes then results in one or more ionized states of the defects, up to the limit set by the ionization potentials

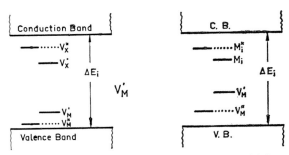

Fig. 1.14. Intrinsic defect equilibria and ionization of defects.

and electron affinities of the species in the crystal. A neutral anion vacancy constitutes a donor level, in most cases lying just below the conduction band (Fig. 1.14); a neutral cation vacancy an acceptor level just above the uppermost filled valence band. Similarly, interstitial metal atoms are donors, interstitial nonmetal atoms acceptors. In highly ionic compounds the neutral point defect states may lie within the conduction and valence bands respectively, so that the first ionized state is the ground state.

On this basis, the defect equilibria (1.3) to (1.5) must be replaced by a set of quasi-chemical equations which includes the valence band/conduction band electronic equilibrium and the ionization of neutral defects. Thus thus for a compound $A^{2+}B^{2-}$

$$O \rightleftharpoons V_A^x + V_B^x \qquad H_s \quad \text{Equilib. const. } K_s \qquad (1.17)$$

$$O \rightleftharpoons e_{vb}^+ + e_{cb}^- \qquad E_i \qquad\qquad K_i \qquad (1.18)$$

$$V_A^x \rightleftharpoons V_A' + e^+ \qquad E_a \qquad\qquad K_a \qquad (1.19)$$

$$V'_A \rightleftharpoons V''_A + e^+ \qquad\qquad E_{a2} \qquad\qquad K_{a2} \qquad\qquad (1.20)$$

$$V^x_B \rightleftharpoons V^{\cdot}_B + e^- \qquad\qquad E_b \qquad\qquad K_b \qquad\qquad (1.21)$$

$$V^{\cdot}_B \rightleftharpoons V^{\cdot\cdot}_B + e^- \qquad\qquad E_{b2} \qquad\qquad K_{b2} \qquad\qquad (1.22)$$

$$0 \rightleftharpoons V'_A + V^{\cdot}_B \qquad H'_s = H_s - (E_i - E_a - E_b) \quad K'_s \qquad (1.17a)$$

$$0 \rightleftharpoons V''_A + V^{\cdot\cdot}_B \qquad\qquad H''_s \qquad\qquad K''_s \qquad\qquad (1.17b)$$

with the constraint for electrostatic charge balance

$$[V^{\cdot}_B] + 2[V^{\cdot\cdot}_B] + n = [V'_A] + 2[V''_A] + p \qquad\qquad (1.23)$$

and the nonstoichiometric transfer reaction in the form

$$A(g) \rightleftharpoons A^x_A + V^x_B \quad \text{or} \quad A(g) \rightleftharpoons A^x_I \qquad\qquad (1.24)$$

$$\tfrac{1}{2}B_2(g) \rightleftharpoons V^x_A + B^x_B \quad \text{or} \quad \tfrac{1}{2}B_2(g) \rightleftharpoons B^x_I \,. \qquad (1.24a)$$

This set of equilibria can be, and most often is, handled empirically, fitting the energies of reactions (1.17)–(1.22) and their equilibrium constants to experimental data. It can, however, be treated more rigorously and Brebick [1958, 1961] has evaluated the grand partition function for the point defect and electronic equilibria of a nonstoichiometric crystal in terms of the band model. In considering the implications of the elaborated point defect approach it is convenient, following Kröger, to build up the equilibrium diagram by considering alternative extreme simplifications of the electroneutrality condition in different regimes. For simplicity we consider systems of the general type AB_{1+x}, in which only a single charge state is important for either type of native defect. For a high metal-excess composition, the approximation $n = [V^{\cdot}_B]$ can be made (regime I); for a high nonmetal excess, $p = [V'_A]$ (regime III). At both these limits nonstoichiometry can be treated as involving only one kind of point defect. In the intermediate range, where both conjugate types of defect must be considered, the Kröger-Vink treatment brings out the fact that nonstoichiometric behaviour depends strongly on both the intrinsic disorder and the electronic properties of crystals.

The first possibility, valid for crystals with a wide band gap E_i, is that

$K'_s > K_i$. Then regime II corresponds to $[V'_A] \approx [V'_B]$ (Fig. 1.15). If, as is the case for highly ionic solids, K_a, K_b are relatively large, $K'_s > K_s (= K_a K_b K'_s / K_i)$, the exact stoichiometric point corresponds to $[V'_A] = [V'_B] = K'^{\frac{1}{2}}_s = \xi$. In the equilibrium diagram, plotting $\log[N_j](j = $ defect species) against

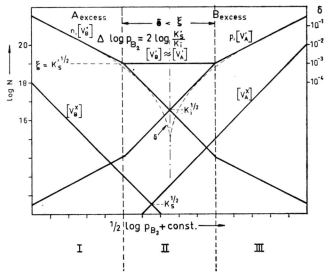

Fig. 1.15. Defect concentration and stoichiometric variation for crystal with $K'_s > K_i$. Logarithmic plot as function of chemical potential of nonmetal.

$\log p^{\frac{1}{2}}_{B_2}$, the stoichiometry of the crystal is very insensitive to a large change in the equilibrium pressures. On the metal-rich side, the limit of regime II may be defined by the condition

$$[V'_A] = n = (1/\sqrt{2})[V'_B] , \quad \delta = -(1/\sqrt{2})K'^{\frac{1}{2}}_s ;$$

the nonmetal-rich limit is given by

$$[V'_B] = p = (1/\sqrt{2})[V'_A] , \quad \delta = +(1/\sqrt{2})K'^{\frac{1}{2}}_s .$$

The corresponding range of chemical potentials is

$$\Delta \mu_{B_2} = 2RT \log (K'_s / K_i) .$$

Hence if the native disorder is small and the band gap wide, a crystal behaves as a compound of analytically constant composition over a very wide range of conditions; only the change in free carrier concentration indicates the minute deviations from ideal stoichiometry. This is the situation for the typical ionic compounds and gives a meaning to the nature of line phases in equilibrium diagrams. Thus, for KBr at 600°C, from the data tabulated by Kröger [1964], $K_s' = 8.4 \times 10^{-14}$, $K_i = 3.3 \times 10^{-35}$, whence $\delta < 10^{-6}$ over a range of 10^{43} in the equilibrium bromine activity.

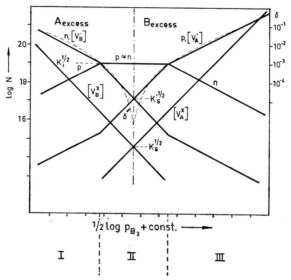

Fig. 1.16. Defect equilibria and stoichiometric variation for crystal with $K_i > K_s'$.

The second possibility (Fig. 1.16) is that, in regime II, $n \approx p$; the electronic properties become insensitive to the composition. Then, in this central regime, $[V_B^{\cdot}] \propto p_{B_2}^{-\frac{1}{2}}$, $[V_A'] \propto p_{B_2}^{\frac{1}{2}}$, and the stoichiometric variation is related to the equilibrium pressure of B_2 vapour by equation (1.16). The central range, in this case, covers a change in chemical potential

$$\Delta \mu_{B_2}''' = 2RT \log (K_i/K_s');$$

at its limits, $\delta = \pm (1/\sqrt{2}) K_i^{\frac{1}{4}}$. For both cases, in regime I $\delta \approx [V_B^{\cdot}]$ and is proportional to $p_{B_2}^{-\frac{1}{4}}$; in regime III $\delta \approx [V_A'] \propto p_{B_2}^{\frac{1}{4}}$. If the defect ionization equilibrium constants K_a, K_b are small, uncharged defects (concentration $\propto p_{B_2}^{\pm\frac{1}{4}}$) may be important in determining the stoichiometric point and the dominant equilibria for relatively large deviations from stoichiometry. Figures 1.17 a,b represent, on a logarithmic scale, the dependence of δ on p_{B_2}; Fig. 1.17a, based on the data for PbS, shows by implication how $\overline{H}_{s_2}$ changes with the composition, according to the enthalpy change in the dominant defect-producing reaction.

A well recognized advantage of the Kröger-Vink approach is that account can readily be taken of subsidiary equilibria, such as minor modes of

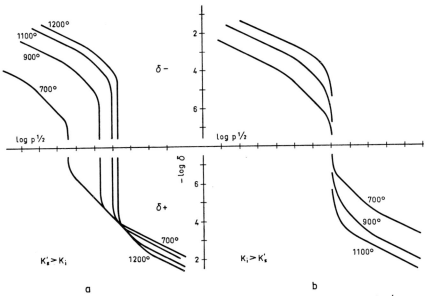

Fig. 1.17. Variation of stoichiometric defect with partial pressure of nonmetal, (a) $K_s' > K_i$; (b) $K_i > K_s'$. Figures for (a) based on PbS. Note small dependence of δ on the temperature in range III.

native disorder and association reactions between defects. It is conceivable, in principle, that the equilibrium constants for association reactions are such that "free" vacancies or interstitials, on which the foregoing diagrams and discussion have been based, are present in much smaller concentration than defects in complexes. Nevertheless, the treatment is essentially limited to dilute regular solutions and its proper area of applicability is to systems with small deviations from stoichiometry. Within that area it provides a good quantitative model for semiconducting behaviour, diffusion processes, stoichiometric variations, the effects of doping with foreign atoms etc. As applied to narrow band gap solids $(K_i > K_s)$ it may well be valid up to the

References p. 74

degree of nonstoichiometry encountered, for example, in SnTe and related compounds (Brebrick [1963]).

However, the relation between the excess chemical potential of a nonstoichiometric compound and the intrinsic degree of imperfection ξ (eq. 1.16)), and the deduction that $\delta \leqslant \xi/\sqrt{2}$ in the central range for wide band gap, ionic solids, makes it difficult to interpret the properties of grossly nonstoichiometric compounds. In a significant number of cases such compounds have been shown, from comparison of measured densities and cell dimensions, to have a high concentration of unfilled lattice sites. Thus for stoichiometric TiO and VO, about 15 per cent of lattice sites are unfilled (i.e. the stoichiometric compounds should be represented as $M_{0.85} O_{0.85}$) and apparent vacancy concentrations of 4–20 per cent of sites in both metal and nonmetal sublattices have been found in transition metal chalcogenides (see Ehrlich [1949], Wadsley [1964]). It is difficult to regard these defect concentrations as arising from simple point defect equilibria (eqs. (1.17), (1.17a), (1.17b)) without arbitrary assumptions about the enthalpy of vacancy creation. Moreover, although it is convenient to draw a distinction between compounds that exhibit the small stoichiometric variability implied by point defect theory and those that are grossly nonstoichiometric, the distinction is arbitrary. As has already been pointed out, the same compound may be stoichiometric, subject to point defect equilibria, at low temperatures, but grossly nonstoichiometric at high temperatures. These considerations lead inescapably to the conclusion that some substantial modifications must be made in extending the completely general and inherently valid ideas of lattice imperfections to that much smaller group of compounds that displays a wide stoichiometric range.

6. Interactions between defects

6.1.

One simplifying assumption, obviously too crude, that has been made in the discussion thus far has been to take the enthalpy of creation of defects as independent of their concentration and their particular spatial arrangement.

At the lowest level, since the Kröger-Vink ionized point defects are charged entities, they are subject to the same kind of electrostatic interactions – Debye-Hückel effects – as govern the distribution of ions in dilute electrolyte solutions. For a mole fraction x_i of defects of each kind, bearing charges z_i, in a crystal of molar volume V and dielectric constant ε, a Debye length can be defined by

$$L_D = \left[\frac{V\varepsilon kT}{4\pi e^2 \Sigma x_i z_i^2} \right]^{\frac{1}{2}} . \qquad (1.25)$$

Then the activity coefficient $\gamma = \exp[-z_i^2 e^2/2\varepsilon kTL_D]$ in the symmetrical case, and the Schottky equilibrium constant becomes

$$K_s' = K_s^0 \exp\left[-\{H_s'/kT - z_i^2 e^2/2\varepsilon kTL_D\}\right] = K_s^0 \exp\left[-H_s^*/kT\right] \qquad (1.26)$$

where $H_s^*(T) < H_s'(T)$. The effect of the ionic atmosphere is therefore to increase the equilibrium concentration of defects, but the effect is small; for NaCl at its melting point, $H_s' \sim 2$ eV, $H_s^* = H_s' - 0.033$ eV. Whilst the equilibria should strictly be interpreted in terms of activities rather than concentrations or mole-fractions, the pure concentration effect expressed by the Debye-Hückel correction has no relevance for the present problem.

Of far greater importance is the neglect of configurational effects. Denoting by N_D the total number of defects of all kinds in a crystal of N atoms, and by $\{N_D\}$ a particular configuration of those defects, the total free energy of the crystal should be represented as the sum of three terms:

$$F = F_0(N) + F(N, N_D) + F(\{N_D\}) \qquad (1.27)$$

where $F_0(N)$ represents the free energy of the perfect (i.e. stoichiometric and fully ordered) crystal, $F(N, N_D)$ is that part of the free energy attributable to the introduction of N_D defects (including, for a nonstoichiometric crystal, the energies of valence changes) and $F(\{N_D\})$ is a configurational free energy. Some attempt can be made to include the contributions to the configurational free energy from defect association reactions, but a general solution would require that the interaction of all combinations of pairs, triplets etc. should be summed over the whole crystal:

$$F_{config} = \Sigma_{ij} N_{ij} E_{ij} + \Sigma_{ijk} N_{ijk} E_{ijk} + \dots .$$

Even for dilute defect systems (Allnatt and Cohen [1964]) the solution of such a problem in cluster theory is a formidable task, and yields results that are still valid only in the low concentration range that is applicable to pro-

References p. 74

blems of F-centres and doping in alkali halides etc. The attempts that have been made to include the configurational free energy in the statistical mechanics of chemically interesting nonstoichiometric compounds have therefore been confined to the treatment, in different ways, of pair-wise interactions.

6.2. Attractive interactions between defects

The most direct approach to inclusion of the configurational free energy is to take account of the interaction energy of defects on nearest neighbour sites. This may be a poor approximation for ionic solids but is justifiable as an extension of the Schottky-Wagner model, in which localization of charges is ignored. This procedure, an extension of Fowler's statistical mechanical treatment of localized monolayers, was first applied by Lacher [1937] to the classical nonstoichiometric palladium-hydrogen system, and was extended by Anderson [1945] and Rees [1954] to provide a general theory of the existence ranges of binary compounds, since it gives an interpretation of the critical miscibility phenomena observed in many systems and the more general intuitive view that any crystal must have a limited tolerance for defects.

Consider a crystal of the compound AB, with $2N$ lattice sites, incorporating N_i defects of type i. These may be interstitials or vacancies, and if each defect site is surrounded by z_i nearest neighbour sites of the same kind the number of nearest neighbour defect pairs, for a random distribution, is $z_i N^2/2N$. The interaction energy between neighbouring defects may be represented as $2E_{ii}/z_i$ and is zero between defects that are not nearest neighbours. Then the configurational free energy $F(\{N_D\})$ (eq. (1.26)) may be represented by

$$F(\{N_D\}) = \sum_i - N_i^2 E_{ii}/NkT \tag{1.28}$$

in which the N_i are the equilibrium values for the defect concentrations.

If, in particular, we consider a crystal with Frenkel type native disorder in the metal sublattice, in equilibrium with the nonmetal vapour B_2, its properties on the metal-excess side of the ideal composition will be modified by the interaction E_{II} between interstitials, and on the nonmetal-excess side by E_{VV}, the interaction between vacancies. For crystals with a substantial stoichiometric defect it is convenient to introduce the quantities $\theta_I = N_I/N$, $\theta_V = N_V/N$. It is then found that the equilibrium condition is given by

$$p_{B_2}^{\frac{1}{2}} = \frac{1}{g(T)\cdot\kappa(T)} \frac{1}{1-\theta_V} \exp\left[-\frac{E_{XB} - 2\theta_V E_{VV}}{kT} \right] \tag{1.29}$$

$$p_{B_2}^{\frac{1}{2}} = \frac{1}{g(T)\cdot\kappa(T)} \frac{1-\theta_I}{\theta_I} \exp\left[-\frac{E_{VA} + E_{IA} + E_{XB} + 2\theta_I E_{II}}{kT} \right] \tag{1.30}$$

where $g(T)$ comes from the partition function of the diatomic molecule B_2 and other symbols have the meanings assigned in eqs. (1.12) etc. Except for compositions close to the ideal (for which $\delta = \theta_V - \theta_I$), the form of the $p-X$ isotherms usually depends on changes in concentration of one defect type only, so that it is permissible to consider (1.29) and (1.30) separately. Then, on the nonmetal-excess side, by analogy with (1.16), we have

$$\left(\frac{p(\theta)}{p(\frac{1}{2})}\right)^{\frac{1}{2}} = \frac{\theta_V}{1-\theta_V} \exp\left[-\frac{(1-2\theta_V)E_{VV}}{kT} \right] \tag{1.31}$$

with a similar expression in θ_I, E_{II} on the metal-excess side. Provided that E_{VV}, $E_{II} < 0$ (i.e. if like defects effectively attract one another), (1.31) and its analogue have the important property that, at high temperatures, θ is a single valued function of $p(\theta)/p(\frac{1}{2})$ – the composition is a continuous function of

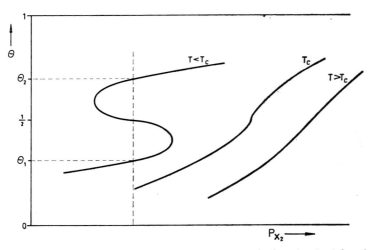

Fig. 1.18. Defects-with-interaction model (Anderson, Rees). θ as 3-valued function of equilibrium pressure below T_c.

References p. 74

the equilibrium pressure from $\theta=0$ to $\theta=1$ – but, at low temperatures, there is some value of $p(\theta)/p(\frac{1}{2})$ that is satisfied by three values of θ ($\theta=\frac{1}{2}$ and two other values, θ_1, θ_2) (Fig. 1.18). The physical significance of this is that the solution θ_1 represents the saturation concentration of defects in the phase AB, θ_2 the saturation composition of the phase that coexists with the limiting AB. This can alternatively be described as resulting from a discontinuous enrichment of defects, or as the lower stable composition of a phase in which the defects of AB have become structure elements. As the temperature rises, the coexisting compositions θ_1 and θ_2 approach each other, and the two-phase region disappears at a temperature $T_c = -2E_{ii}/k$. On this basis, the absolute stable range of a compounds phase is determined by the energies of interaction of the types of defect that are implicated in non-stoichiometric properties on either side of the ideal composition.

A negative interaction energy E_{ii} implies that defects tend to cluster. Such clustering of interstitials represents the progressive occupation of a sublattice which is empty in the compound AB; saturation and phase separation represent the formation of a topologically related structure in which this newly (but incompletely) occupied sublattice is an integral part of the ordered structure. Care is required in extending the same considerations to the clustering of vacancies and formation of a new structure with a high concentration of unoccupied sites, which are structure elements, not defects. The effect of going from a structure AB to a structure AB_n ($n>1$), formally related to AB by an ordered arrangement of vacant A sites, is to change the coordination of A atoms about B atoms. In so far as this arises from the ordering of A vacancies, isolated vacancy pairs or clusters may be defined by their mutual orientation, as well as their position, and an additional orientation entropy should then be included.

A further nett effect of clustering is to stabilize configurations with defects, so that the degree of native disorder in the stoichiometric crystal is increased. In place of (1.16) we find

$$\frac{\xi}{1-\xi} = \exp\left[-\frac{(E_{VA}+E_{IA}+2\xi E_{ii})}{2kT}\right] . \tag{1.32}$$

The effect of the interaction energy term on the native disorder is not large, and cannot by itself account for the apparent high degree of native disorder in such compounds as TiO or CrS, referred to previously.

Because electronic terms and defect ionization states are not included in the theory, it can no longer be expected to give a satisfactory interpretation of nonstoichiometric ionic solids with K_s' or $K_f' > K_i$. It may, however, be applicable to systems with $K_i' > K_s'$ or K_f, especially for those in which the mobility of the defective sublattice is high, so that statistical disorder is maintained. There is a lack of chemical potential/composition data for the highly nonstoichiometric carbide systems, but the quasimetallic hydrides of the transition metals and lanthanides have been considered by Libowitz [1962, 1963], who has also extended the treatment to include Schottky disorder; this is formally more complex since lattice sites are not conserved. In making a comparison between the theoretical expression and experimental data (which has at least a heuristic value) E_i and E_{ii} are the only disposable parameters that need to be determined empirically. These have been considered above as independent of the temperature, and no reference has been made to the influence of $\kappa(T)$, the contribution of hyperstoichiometric atoms to the vibrational partition function. This can be handled in several ways (see Thorn and Winslow [1966], Atlas [1968]), and can be shown to have the effect of making E_i and E_{ii} temperature dependent quantities. If this factor can be handled explicit, it suffices to fit the equilibrium expression to any two experimental points. From the values thereby assigned to E_i, E_{ii}, the complete equilibrium diagram should be reproduced, including the $(p–X)_T$ isotherms, the phase limits as a function of temperature and the critical miscibility temperature T_c. Hoch [1962, 1963 a, b] has sought to apply the next neighbour interaction model to such systems as TiO, which must, over the whole composition range, be treated as containing two defect types. It is then necessary to introduce a correspondingly larger number of interaction energy parameters as empirically determined quantities, and although the values found are reasonable, the number of disposable parameters detracts from the significance of the results.

Two examples may be cited of the application of the theory to real systems. The first concerns the lanthanide hydrides, for which (in view of their metallic properties) a treatment in terms of atomic defects may be appropriate. Hydrogen enters tetrahedral sites in the f.c.c. array of metal atoms, to form the hydrides MH_2, with a narrow substoichiometric range of composition. The MH_2 phase takes up further hydrogen, in octahedral sites, to the upper limiting composition MH_3 when all octahedral sites are filled. Numerous studies of equilibria have been made in this nonstoichiometric range, and n.m.r. measurements (e.g. Schreiber and Cotts [1963]) have shown that proton place exchange takes place freely at room temperature.

References p. 74

Messer and Hung [1968] have recently analysed data for LaH_n ($2.2 < n < 2.7$) in terms of a regular solution model which, in a slightly different formalism, corresponds exactly to that outlined above. The nonstoichiometric LaH_n can be regarded as derived from the fluorite structure of LaH_2 by occupation of interstitial sites, or from that of LaH_3 by creation of vacancies in the octahedral sublattice. It is tacitly assumed that the tetrahedral sublattice remains fully populated at all compositions and temperatures, and this assumption is probably unjustified. The interaction energy therefore needs careful definition, since "defective" and "proper" occupation of octahedral sites are interchangeable terms. Messer and Hung therefore consider vacancy–vacancy, interstitial–interstitial and vacancy–interstitial interactions, and define a nett interaction energy: $w = 2E_{VI} - E_{VV} - E_{II}$. This is the same nearest neighbour site preference energy as is used in the treatment of short range ordering. We may define $\theta = n - 2$, since only the distribution of hydrogen over octahedral sites is of interest. Then, taking LaH_2 as the reference state and evaluating the activity coefficient of hydrogen from regular solution theory, Messer and Hung obtain an expression which can be transformed into (1.31), but with E_{ii} replaced by the more precisely defined site preference energy w. When w is evaluated at each temperature from the $(p - X)_T$ isotherms, it is found that the theoretical expression reproduces the behaviour of the system well, but that w varies regularly with the temperature, from -1800 cal/mol at $250°C$ to -1080 cal/mol at $450°$. These values are quite comparable with those deduced for other hydride systems (see, for example, Libowitz [1963]), and the temperature-dependence of w, which appears in the regular solution treatment as an excess entropy of mixing, is associated (probably rightly, in part at least) with the vibrational entropy of the rather mobile hydrogen atoms.

A second system to which the nearest neighbour interaction model has been applied in detail is the $UO_{2 \pm x}$ system (Thorn and Winslow [1966]). For such an essentially ionic compound, the omission of defect ionization effects must make the foregoing treatment incomplete. On the substoichiometric side, nonstoichiometry can be discussed in terms of oxygen vacancies. $UO_{2 + x}$ is treated by Thorn and Winslow in terms of interstitial oxygen atoms, although it is known that local rearrangement produces the Willis defect complex in place of an oxygen atom on an octahedral site. As long as these complexes are independent units, the main effect of this simplification is probably to omit an orientation entropy term from the configurational entropy. The value of the analysis lies partly in the explicit treatment of the vibrational entropy change due to defects, which is shown to have the effect of making the defect energies vary with temperature. Stripp and Kirkwood [1954] discussed the vibrational partition function of a closepacked monatomic crystal containing vacancies, and showed that the total partition function can be resolved into a static part Q_s and a vibrational part Q_v. If V_{aa} is the static binding potential between a pair of atoms on next-neighbour sites, in a crystal with n vacancies and N atoms, on $S = n + N$ sites, Q_s was found to be:

$$Q_s = \exp\left[-\frac{V_{aa}}{kT}\left\{\frac{(N-n)z}{2} + P_{aa}\right\}\right]$$

$$= \exp\left[-\frac{SzV_{aa}}{2kT}\right]\exp\left[\frac{nzV_{aa}}{kT} - \frac{P_{aa}V_{aa}}{kT}\right] \tag{1.33}$$

where P_{aa} is the number of nearest neighbour vacancy pairs in a particular configuration. Except that (1.33) makes no distinction between the energy required to create a vacancy in a perfect region and in a defect-rich region (i.e. it omits the energy change designated as the interaction energy), Thorn and Winslow show that (1.33) is equivalent to the static part of (1.31). For the vibrational part Q_v, in terms of the vibrational partition function $Q_v^0(N)$ of the perfect crystal, it is found that

$$Q_v = Q_v^0(N) + 1.725n - 0.2442P_{aa}. \tag{1.34}$$

Taking this approach as applicable to the more complex case of UO_2, and with $z = 6$ for the coordination of octahedral sites, Thorn and Winslow obtain for the contribution of vacancies to the vibrational partition function

$$\ln q(N_v) = 1.725N_v - 0.7326(N_v^2/2N_u) - N_v\ln q_v \tag{1.35}$$

(N_u = number of uranium atoms, determining all site numbers), in which the last term can be quantified by suitable choice of a Debye temperature Θ_v for vacancies. Including both static and vibrational parts, (1.31) then becomes

$$\tfrac{1}{2}\ln p_{O_2} = \ln\frac{1-\theta_v}{\theta_v} - \left(\frac{E_v}{kT} - 1.725\right)$$

$$+ 2\theta_v\left(\frac{E_{vv}}{kT} - 0.7326\right) - \ln q_v - \ln g(T). \tag{1.36}$$

As in eqs. (1.29) and (1.30), $g(T)$ comes from the partition function of O_2 and the conversion of absolute activities into measured pressures. In terms of the modified expression (1.36), the critical temperature T_c is given by

$$2.7326 \ T_c = E_{vv}/k.$$

In evaluating the experimental data, E_{vv} could be derived in two ways, using a Debye temperature Θ_v based with some confidence on the lattice dynamics of fluorite structures. The shape of the lower phase boundary suggests that T_c is not very different from the $U + UO_{2-x}$ eutectic temperature, 3023°K, whence $E_{vv} = 16.47$ kcal/mol. Alternatively, mass spectrometric measurements by Ackermann, Chandrasekariah and Rauh give the oxygen atom activities at the lower phase boundary directly (i.e., $p(\frac{1}{2})$ in (1.29)). From these, for $UO_{1.88}$ at 2200°, $E_{vv} = 16.87$ kcal/mol and $E_v = 209.7$ kcal/mol. With these figures, the lower phase boundary is not badly predicted between 1800° and 2500°K.

Comparison of the theoretical and experimental results on the oxygen-rich side is more difficult, because although there is a critical miscibility temperature at 1400°K, the merging of the UO_{2+x} and U_4O_9 phases suggests that the structure is "filled" with one interstitial site per 4 atoms of uranium, whereas at low interstitial concentrations all interstitial sites – 1 per atom of U – are equally accessible. The number of interstitial sites thus depends in some way on the composition. It is this factor that probably determines the unusual shape of the 2-phase upper boundary curve (Fig. 1.2). From the critical miscibility temperature, E_{II} is found as 5.56 kcal/mol. From the experimentally measured oxygen activities, and with an averaged Debye temperature for interstitials (which must be supposed to include any modification in local vibrational modes), the results for $0.025 < x < 0.08$ ($\alpha = 1$) are fitted by $E_{II} = 5.03$ kcal/mol, $E_I = 94.0$ kcal/mol. From $0.16 < x < 0.24$, the predicted upper phase boundary is consistent with experiment if $\alpha = \frac{1}{4}$. Quite apart from the over-simplification of identifying the excess of oxygen with atomic interstitials, no model based on a constant value for the number of interstitial sites per uranium atom appears capable of reproducing the experimental data consistently over the entire composition range. It appears, rather, that above a certain population of interstitial atoms or clusters, some site exclusion principle, related to the ordering pattern of the intermediate phase U_4O_9, operates to restrict the configurational degeneracy.

Essentially the same way of introducing nearest neighbour interactions was used by Rees [1954] in what is the most general approach yet made to a statistical theory of binary compounds. Rees regards binary phases as derived, in a formal sense, from the introduction of B atoms into interstitial sites in the structure of the parent element A. Where the Bravais lattice of A is conserved, save for minor dimensional changes (as in the B1 compounds and other f.c.c. compounds of the f.c.c. elements), this formal relationship is obvious. Where there is a change of symmetry, the relation is not so clear, but the conception could be retained conceptually and thermodynamically by deriving the compound from a (metastable) modification of A. Topological transformations between some of the main structure types of binary compounds have been discussed by Hyde. Within the base structure A, there is a multiplicity of alternative sublattices for possible occupation by B atoms, and occupation of any one of these, B_1, generates new sets of interstitial positions, B_2, B_3, B_4,..., which will differ in their binding energy. The simplest approximation (which may be invalid in many cases) is that the number of B_2 sites available for occupation is equal to the number of B_1 sites occupied, and similarly for site sets of higher order. Then, with $N_0 =$ number of interstitial sites in the set B_1, N_1, N_2,... the number of B atoms in the sublattice sets B_1, B_2,...; E_1, E_2,... the energy change for each atom entering these sites, and E_{11}, E_{22},... the nearest neighbour interaction energies between atoms in the same sublattice set, the partition function of the crystal can be written in the form

$$Q\{N,(N_1+N_2+ \ldots),T\} = Q(N,N_1,T) \times Q(N_1,N_2,T) \times \ldots .$$

Equilibrium is established for the occupation of each sublattice set under the same chemical potential, to form a crystal AB_x, where $x = (N_1 + N_2 + \ldots)/N$. For a 2-sublattice case, such as the Zr–H system originally discussed by Martin and Rees [1954], or the LaH_x system considered above, the resulting expression analogous to (1.31) is:

$$p_{O_2}^{\frac{1}{2}} = \frac{\theta_1}{1-\theta_1} G_1(T) \exp - \frac{E_1 - 2\theta_1 E_{11}}{kT}$$

$$= \frac{\theta_2}{\theta_1 - 2} G_2(T) \exp - \frac{E_2 - 2\theta_2 E_{22}}{kT} \qquad (1.37)$$

where $\theta_1 = N_1/N$, $\theta_2 = N_2/N$ and $G_1(T)$, $G_2(T)$ contain the vibrational partition functions of B atoms in the sublattices 1 and 2, and that of the B_2 molecule. The relevance of this model is that it takes account of disorder in the B sublattice for crystals that are still substantially substoichiometric as compared with the composition reached by saturating the most favourable

set of sites – e.g., for distribution of H atoms between tetrahedral and octahedral sites below the composition MH_2 (Fig. 1.19).

6.3. Repulsive interactions between defects

The alternative postulate is that, in virtue of lattice strain, coulombic
forces etc., the interaction between like defects is repulsive, so that they tend
to establish as great a relative separation as possible within the crystal.
Superstructure ordering then originates from the minimization of the repulsive interaction energy.

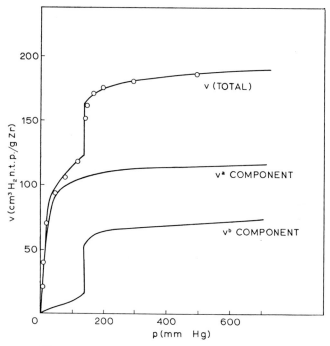

Fig. 1.19. Application of Rees model to ZrH_x system.

In its simplest and most extreme form, this can be treated as a site
exclusion problem: e.g. the presence of an interstitial atom on a particular
lattice site precludes the occupation of some specified number of adjacent
interstitial sites. The energy change for occupation of all other sites is unchanged, however. In effect, the configurational entropy is diminished. This

case has been considered for nonstoichiometric interstitial phases by Speiser and Spretnak [1955].

In the nonstoichiometric crystal $AB_{n+\delta}$, containing N_A atoms of A, the structure provides α sites per atom of A to accomodate the stoichiometric excess of B. If β' interstitial sites are excluded for each site occupied by a B atom, the crystal would reach its saturation capacity with $\alpha N_A/\beta$ interstitial B atoms, where $\beta = \beta' + 1$. The statistical weight of configurations with $N_B(=\delta N_A)$ interstitial atoms is

$$(N_B) = \frac{\alpha N_A(\alpha N_A - \beta)(\alpha N_A - 2\beta)...(\alpha N_A - \{N_B - 1\}\beta)}{N_B!}$$

$$= \frac{\beta^{N_B}(\alpha/\beta N_A)!}{N_B!(\alpha/\beta N_A - N_B)!} . \tag{1.38}$$

This expression for the microcanonical term then replaces the simple combinatorial term in the partition function. If no sites are excluded ($\beta' = 0$, $\beta = 1$), (1.38) reduces to the familiar form. A variant of this approach was tested by Hagemark (Hagemark and Broli [1966]) to interpret the equilibrium data for UO_{2+x}, on the assumption that each interstitial oxygen excludes 12 adjacent sites, so that a mean number of interstitial sites per uranium atom can be defined by $\bar{\alpha} = 1/(1 + 12x)$, to give saturation at the composition $UO_{2.25}$.

A more sophisticated model has been developed by Atlas [1968] and specifically applied to anion-deficient and anion-excess fluorite phases. It merits close consideration because it is very general and does take account of defect ionization equilibria. It involves some ingenious assumptions and approximations, but is perhaps the most promising approach yet made to a statistical mechanical model.

As in the models discussed above, the total energy change in forming a nonstoichiometric crystal is separated into two parts. The first part is the energy change E_f in the transfer of one atom of the anionic component from crystal to gas phase, or vice versa, to produce an ionized defect and its counter ions, at extreme dilution. This is taken as constant. In the fluorite oxides CeO_{2-x}, UO_{2+x}, the defects may be point defects or clusters, and bear an effective charge of z_e; they are balanced electrostatically by altervalent cations with an effective charge of z_e'. These charges are defined relative to the normal components of the crystal and could, in principle, be fractional, mean charges. For simplicity, CeO_{2-x} is regarded as having singly ionized vacancies ($z = 1$), balanced by Ce^{3+} ions ($z' = -1$); UO_{2+x} as having Willis defect clusters with a nett local charge $z = -1$, balanced by U^{5+} ions ($z' = +1$).

The second component of the energy per defect, dependent on composition and configuration, is treated as an interaction energy arising from the balance of Coulomb forces between all like and unlike defect pairs. For any one defect, the total contribution of Coulomb energies to its energy of formation clearly depends on (a) the magnitude of the charges z, z'; (b) the number of other defects and altervalent ions within some volume element which includes all the charged entities that make a significant contribution to the interaction energy; and (c) the distance between the defect in question and every interacting defect.

The central idea is that, as interatomic distances in a crystal are discontinuous, not continuous, functions of position, so the Coulomb interactions can be assigned a set of discrete values. In the oxygen deficient CeO_{2-x}, vacancy–vacancy repulsion energies take on a set of values u_0, u_1, u_2, ..., u_{Lv}, where u_0 is the smallest significant and u_{Lv} the largest permissible value to be included. Consider an isolated pair of vacancies, separated by a distance d_0, such that their repulsion energy is u_0. Taking one of them as centre, the second could be located on any lattice site lying on a sphere of radius d_0, which measures the *interaction volume* for the energy level u_0. Alternatively, to each vacancy there could be ascribed an *envelope volume*, containing C_A anion sites, and with a radius $r_0 = \frac{1}{2} d_0$. For any vacancy closer to the one taken as centre, the mutual repulsion energy u_i would be greater, and measured by the size of the corresponding (smaller) osculating envelope volumes. The discrete values of u_i are then fixed by assigning $C_A - i$ sites to the ith envelope volume.

In the crystal, every vacancy is surrounded by an unknown number of other vacancies and counter-ions at various distances, and the nett repulsion energy is the sum of their component contributions. The problem then is to assign to each vacancy a single vacancy–vacancy repulsion energy. This is achieved by replacing the actual distribution of vacancies and counter-ions by a hypothetical "transformed distribution" of the same total energy. Consider an initially more or less random state. Within some small element of volume around a particular vacancy, let a local fluctuation – e.g. some partial ordering – produce a small local excess of vacancies and reduced cations, that lowers the energy within this region by a small amount ΔE. The same lowering of configuration energy could be achieved by smearing out the cation charges uniformly over all cation sites and decreasing the number of surrounding vacancies. In particular, some transformation could be found in which the total effect of all vacancies was replaced by that of a single vacancy producing a repulsion energy u_i. If each vacancy in turn is taken as the centre of such a transformed distribution, each can be assigned a nett interaction energy and may be said to control a corresponding interaction volume within which no other vacancy can be located. The total number of vacancies in the crystal (i.e. the stoichiometric defect) will then be the sum of the populations $n_0, n_1, n_2, \ldots, n_i$ in each of the interaction energy levels $u_0, u_1, u_2, \ldots, u_i$, and the configurational entropy will be given by the number of ways that their interaction envelopes, with $C_A, C_A - 1, C_A - 2, \ldots, C_A - i$ sites can be distributed over all the anion sites in the crystal. If, for a particular total composition, the equilibrium configuration is one with a high degree of order, a correspondingly large number of vacancies will belong to some particular population group, and the effect will show in a decrease in the configurational entropy.

From the definition of the envelope volume, a crystal with $N_A (= 2N_M)$ anion sites could accommodate a total of N_A/C_A vacancies in the lowest energy state u_0. For the actual population of n_0 vacancies in state u_0,

$$\Omega_{V(0)} = \frac{C_A^{n_0}(N_A/C_A)!}{n_0!\,(N_A/C_A - n_0)!} \,. \tag{1.39}$$

The number of ways of distributing the population n_i of state i is restricted by the condition that the sites available are only those not already included within the envelope volumes of the populations of lower repulsive energy. The number of available sites for vacancies in the ith level is

$$w_i = \frac{N_A + (C_A - i)n_i - \overset{i}{\underset{0}{\Sigma}}(C_A - j)n_j}{C_A - i} \tag{1.40}$$

and the number of configurations for the ith level becomes

$$\Omega_{V(i)} = \frac{(C_A - i)^{n_i}w_i!}{n_i!\,(w_i - n_i)!} \,. \tag{1.41}$$

The total configurational degeneracy for vacancies is then

$$\Omega_V = \overset{Lv}{\underset{0}{\prod}} \Omega_{V(i)} \,.$$

Exactly similar reasoning can be applied to the distribution of counter-ions, which mutually repel with the discrete energies ε_1, ε_2,..., ε_i, and are distributed between the various states with the populations m_0, m_1,...,m_i. Their configurational degeneracy is found as above, and the total degeneracy is $\Omega_{ve} = \Omega_v \cdot \Omega_e$.

Since the interaction energies considered are Coulomb repulsions, the magnitude of u_i can be worked out in terms of the dimensions of an equivalent spherical shell containing $C_A - i$ anion sites. For a fluorite crystal (cell dimension a_0, dielectric constant K_0), this is found as

$$u_i = \frac{z^2 e^2}{2K_0 a_0} \left(\frac{4\pi}{3(C_A - i)}\right)^{\frac{1}{3}} \left(\frac{7C_A + i}{C_A - i}\right) = \left(\frac{C_A}{C_A - i}\right)^{\frac{4}{3}} u_0 . \tag{1.42}$$

If the same values are taken for the minimum significant vacancy–vacancy and cation–cation repulsion energies u_0 and ε_0, the latter can also be written in terms of u_0.

The attractive Coulomb energy between vacancies and counter-ions involves the number of counter-ions in the interaction volume of each vacancy, and this is found from the transformed distribution to furnish a total of $N_{ve} = (4N_v^2 C_A/N_M)(z'/z)^2$ pairs, with a mean attractive energy of $\bar{E}_{ve}$ per pair, which could, in principle, be calculated approximately, or can be regarded as a parameter to be obtained from the experimental data.

Using the reasoning outlined, it is possible to put the rather complex partition function in a form that can be handled numerically. It must first be maximized with respect to the populations in the different energy levels and then, by differentiation with respect to the defect concentration, the composition of the nonstoichiometric phase is obtained as an explicit function of the oxygen pressure. In doing so, as in similar solid state problems, there is some arbitrariness in the value taken for the dielectric constant, which appears as a scaling factor in the Coulomb energies, but there is no other completely disposable parameter except the size of the largest and smallest envelope constants to be considered – i.e. the range of interaction energies included. These are governed largely by the computational effort required; they determine also the range of compositions over which the free energy equation is solved and the significance of the values found for thermodynamic functions in the sensitive portions of the $\bar{G}-x$ etc. curves. Atlas took C_A, the envelope constant for the smallest interaction energy, as including 60 sites, running down to C_{Lv} with 4 sites; this covered the composition range $0.033 < x < 0.5$ in CeO_{2-x}. The contribution from the change in vibrational

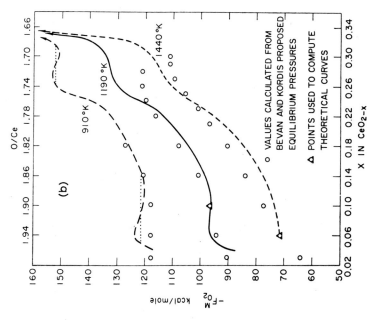

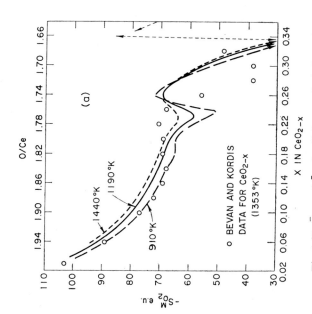

Fig. 1.20. $\bar{F}_{O_2}$ and $\bar{S}_{O_2}$ for CeO_{2-x}: calculations of Atlas model compared with experimental data (Bevan and Kordis).

partition function could either be assumed to be negligible, or included by calculating the change in Debye temperature from the change in cell dimensions of the nonstoichiometric crystal, by means of a Grüneisen relation.

The only quantities not pre-determined are then E_f, the energy of defect creation, and E_{ve}, the mean vacancy–counter–ion attraction energy. These were obtained by solving the free energy equation for any two compositions for which the chemical potential of oxygen was known.

Atlas has compared his model in this way with the experimental results of Bevan and Kordis [1964] for CeO_{2-x} (Fig. 1.20). In the composition range $CeO_{1.97}$ to $CeO_{1.80}$, corresponding to the nonstoichiometric α-phase above 900°K, the thermodynamic properties are reproduced fairly well. For $x > 0.2$ rather poorer agreement is attributable, in part, to arithmetical limitations in the calculation of configurational entropy. A significant feature, however, is that the thermodynamic functions, and especially $\bar{S}_{O_2}$, show instabilities at just those compositions for which, below 900°K, 2-phase fields occur. These fluctuations in the calculated values persist at higher temperatures, however, into the 1-phase region, and this is taken as indicative of a considerable degree of clustering and statistical order for these same average defect densities, even above the critical miscibility temperature.

For CeO_{2-x}, the defects were treated as vacancies occupying single lattice sites. In extending his model to UO_{2+x}, Atlas based the configurational entropy on the Willis 2:1:2 cluster, with two anion vacancies and interstitial atoms on two kinds of site, O' and O''. Assuming that each cluster traps one U^{5+} counter-ion on one of the immediately adjacent cation sites, the nett charge per cluster is -1, balanced by U^{5+} ions distributed in the surrounding crystal structure. Since the Willis cluster does not have cubic symmetry, the configurational entropy must contain an additional factor to allow for the possible orientations of the clusters. As may be seen (Fig. 1.21), the experimental data are reasonably well reproduced; $\bar{S}_{O_2}$ and $\bar{F}_{O_2}$ again display breaks or instabilities at the boundaries of the $UO_{2+x} - U_4O_{9-y}$ 2-phase field. Above 1400°K, where the upper boundary of UO_{2+x} extends to the composition $UO_{2.25}$, these instabilities are no longer catastrophic, but still persist. They could, again, be taken as indicative of a considerable measure of local order. In effect, the Atlas model optimises the interaction energy on a radial distribution function; for appropriate compositions this passes over into an ordered state at lower temperatures and persistently retains relicts of the local structure of that state, with a predominant defect–defect spacing, even when long range correlation has been lost.

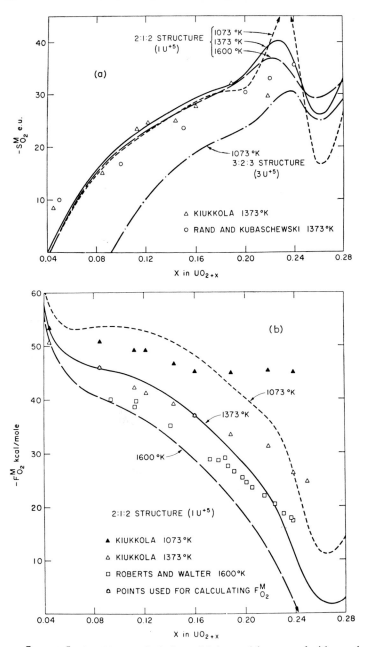

Fig. 1.21. $\bar{F}_{O_2}$ and $\bar{S}_{O_2}$ for UO_{2+x}: calculations of Atlas model compared with experimental data.

7. Random defects in quasimetallic compounds

Theoretical work and crystallographic studies of ordered intermediate phases have been largely concentrated on ionic compounds in which the relatively long range Coulomb forces are probably the most important factor in establishing order. Less attention has been paid to the defect chemistry (as distinct from the metallurgical properties) of the highly nonstoichiometric phases common amongst the carbides, nitrides, borides, etc. of the transition elements; the large group of phases with structures of or derived from the B1 type is of particular interest. There are certain clear tends in the stoichiometric variability of these compounds as the transition metal is varied, and as the nonmetal is changed in the sequence $B \rightarrow C \rightarrow N \rightarrow O$, but thermodynamic information is at present sparse. Nonstoichiometry is associated with very incomplete occupation of the nonmetal sublattice but (as, for example, in the high temperature TiO) may involve high concentrations of vacancies in both sublattices. They appear to be statistically random structures, but this may be illusory in the sense that the high diffusion rates of the nonmetal atoms preclude the existence of anything but extreme short-range ordering. On the other hand, the concentration of both metal and nonmetal vacancies in TiO and VO, for example, is so high that a simple random statistics calculation of nearest neighbour site occupancy would indicate that a considerable fraction of the atoms would have very improbable environments (see Anderson [1964]). It is difficult to escape the conclusion that there is a considerable degree of correlation between sites, at the short range level, even on an instantenaous time scale. The interaction between defects is certainly smaller than in ionic solids (compare the interaction energies, cited above, for "interstitials" in the hyperstoichiometric fluorite phases – quasimetallic LaH_{2+x} and ionic UO_{2+x}), but nevertheless significant. Thus the carbides and nitrides of the later transition groups, and the lower carbides of the lanthanides, form a variety of phases based on ordering over sets of sites of the B1 structure; TiO has a reversible transition to a highly ordered structure (Watanabe, Castles et al. [1967]), which may shed some light on short range order in the high temperature B1 phase; recent work (e.g. Alyamovskii, Gel'd et al. [1968]) has given good evidence for a series of ordered phases ($V_{32}C_{28}$, V_7C_6) in the highly nonstoichiometric V–C system. A resume of the existing state of knowledge of stoichiometries and structures is given by Goldschmidt [1967].

The essential property of all these materials is that they are *metals*, with a large population of electrons in unfilled and possibly overlapping bands. If, for a particular phase MX_{1-x}, the density of states function is not very sensitive to changes in the stoichiometric defect x, the main contribution to changes in the internal energy of the crystal, as the stoichiometry changes, may come from the entry of electrons into higher energy states in the conduction band. The band structure of the B1 type crystals is determined by the overlap of the *ns, np, nd* and $(n+1)s$ orbitals of the metal atoms, and the $2s$ and $2p$ orbitals of the nonmetal atoms, at the interatomic distances fixed by the observed crystal parameters. The electron affinity of the nonmetal affects the energy of states that are derived, in principle, from bonding orbitals of an MX molecule; atomic states lying above the core states of the metal atoms, which largely determine the nonbonding and antibonding orbitals of the MX molecule, make the main contribution to the uppermost, conduction band.

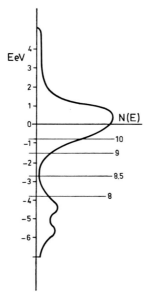

Fig. 1.22. Density of states function for TiC and similar B1-type transition metal compounds (after Bilz), showing variation of Fermi level with electron : atom ratio.

Chemical specificity enters in this way, and in particular, the progressive change in relative energies and radial extension of the cation e_g and t_{2g} orbitals is important. Nevertheless, it is a reasonable first approximation

References p. 74

to consider all the B1 phases as being more or less the same, but differing in the electron : atom ratio. Differences arising from the core potentials of different metal atoms and the electron affinities of the nonmetal atoms can be taken as of secondary importance (Denker [1964, 1968]). One electron calculations of the band structure have been carried out (Bilz [1958] and Denker, loc cit.) for TiC, TiN and TiO, and yield a density of states function (Fig. 1.22) with a considerable band overlap (hence the full metallic character), in which the Fermi level lies well above the uppermost bonding state. Changes in structure or stoichiometry that would decrease the occupation of states in the upper band can therefore lower the free energy per atom in the compound.

Table 1.4

Electron : atom ratios and stoichiometry of some MX compounds

	Valence electrons per metal	Stoichiometric range	Constitution of stoichiometric phase	Electrons per M site in the stoichiometric phase
TiC	8	$TiC_{1.0} - TiC_{<0.5}$	$TiC_{1.0}$?	8
TiN	9	$TiN_{1.2} - TiN_{0.38}$	Probably a vacancy structure	< 9?
TiO	10	$TiO_{1.25} - TiO_{0.7}$	$Ti_{0.85}O_{0.85}$	8.5
VC	9	$VC_{0.96} - VC_{0.75}$	Nonexistent	—
VN	10	$VN_{1.0} - VN_{0.72}$	Probably a vacancy structure	?
VO	11	$VO_{1.3} - VO_{0.7}$	$V_{0.85}O_{0.85}$	9.4

Table 1.4. shows the electron : metal atom ratios for titanium and vanadium carbide, nitride and oxide. In stoichiometric TiC, the valence electron concentration is enough exactly to fill the bonding states; the compound is metallic because of band overlap, but the stoichiometric carbide is stable with a low native vacancy concentration. For all the other compounds, perfect stoichiometry would involve electrons in antibonding states. In so far as the band structure is determined principally by overlap of metal atom orbitals, the internal energy can be lowered by depopulating the uppermost

states, and the configurational entropy increased, by leaving nonmetal sites vacant. The composition ranges with a deficiency of nonmetal, and the non-existence of stoichiometric VC, are qualitatively explicable. However, the nitrides and oxides have compositions up to or exceeding the stoichiometric value; for the oxides it is known, and for the nitrides it is plausible, that the compounds are highly defective, with vacancies on both sublattices. Denker has suggested that lattice vacancies leave unfilled states, lying below the Fermi level of a perfect crystal, so that energy expended in creating vacancies in *both* sublattices is partially offset by filling only the lower-lying levels and depopulating the upper band. Thus, stoichiometric TiO, with 15 per cent cation and anion vacancies, can be regarded as a compound with 8.5 electrons *per cation site*, with a consequential lowering of 1.75 eV in the Fermi energy. On the basis of some rather uncertain defect energies derived by Hoch from the defects with interaction model, it can be calculated that the creation of a $V_{Ti}-V_O$ vacancy pair costs 0.62 eV. Thus there is a nett gain from the formation of the defective structure.

Implicit in this view is the assumption that the parameters determining the band structure are so completely determined by the statistical pseudo-symmetry of the crystal structure that the absence of positive cores from a high proportion of sites does not too drastically perturb the inner potential. This may be an uncertain assumption.

8. Nonstoichiometric compounds as highly ordered systems

8.1.

During the last few years, a number of workers have moved away from the concepts based on disordered structures, towards the view that nonstoichiometric compounds retain a very high degree of local order, certainly going beyond the nearest neighbour level. Taking this view to the extreme limit, the concept of a defect disappears, and the observed crystallographic and thermodynamic properties of nonstoichiometric phases must be attributed entirely to the loss of all long range correlations within a crystal.

It appears to have been first explicitly suggested by Ariya (Ariya and Popov [1962], Ariya and Morozova [1958], Erofeeva, Lukinykh and Ariya [1961]) that, in nonstoichiometric compounds, discrete regions or "islands" having one structure and composition were dispersed in a matrix of another structure and composition. Each region may be ordered and defect-free, but the dispersion is random. Ariya recognized and qualitatively described some features of such a "submicroheterogeneous" system, but did not pursue the thermodynamic implications or relate the hypothesis to the sort of crystallographic and structural evidence considered earlier.

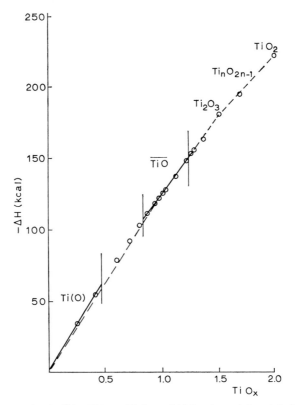

Fig. 1.23. ΔH_f^0 for TiO_{1+x}, Ti_2O_3 and TiO_2 phases (after Ariya).

Ariya's hypothesis was based primarily on thermochemical evidence. He observed that, in the Fe–O, Ti–O and V–O systems, ΔH_f^0 for any preparation MO_x ($1.0 < x < 1.5$) was a linear function of composition:

$$\Delta H_f^0 (MO_{1+y}) = (1-2y) \Delta H_f^0 (MO_{1\cdot 00}) + 2y \Delta H_f^0 (MO_{1\cdot 5})$$

not only for two-phase mixtures of M_2O_3 with the upper limiting monoxide phase, but also within the phase field of the nonstoichiometric monoxide

(Fig. 1.23). In a formal sense, MO_{1+y} behaves as a solution of $MO_{1.5}$ in $MO_{1.00}$, with zero enthalpy of mixing, and Ariya proposed that blocks of the one structure were indeed present as a solute, in the other structure as solvent. This statement is subject to certain restrictions. Since the nonstoichiometric oxides are monophasic in their diffraction properties, any substructure must be on an extremely small scale and completely randomized. The compounds in question are assigned the B1 structure, so that even if the f.c.c. cell is only a pseudo-cell, every atom must occupy a site assignable to the B1 structure. From the thermochemical argument, Ariya deduced that the microdomain structure of TiO_{1+y} was of Ti_2O, with oxygen occupying one half the octahedral sites in a perfect Ti lattice, and of $TiO_{1.5}$, with one third of the octahedral cation sites vacant in a fully occupied anion lattice. As present in TiO_{1+y}, the packing is cubic; both Ti_2O and Ti_2O_3 are based on hexagonal close packing. Formation of the microdomain structure would therefore appear to involve some (necessarily endothermic) transformation, which is at variance with the postulated zero enthalpy of mixing unless some special hypotheses are introduced. Where it can be assumed that the structure of the microdomains are identical with those of the stable end compounds, or are only slightly endothermic, the microdomain idea becomes more plausible. The crucial question is why the highly dispersed, intergrown mixture of two structures should be stable against unmixing into its stable components. This stability implies that

$$\Delta G^0(MX_y) < \{x_A \Delta G^0(MX_a) + (1-x_A) \Delta G^0(MX_b)\} \,,$$

from which, with zero or very small enthalpy change, it can be inferred that

$$\Delta S^0(MX_y) > \{x_A \Delta S^0(MX_a) + (1-x_A) \Delta S^0(MX_b)\} \,.$$

The validity of attributing this relation to an entropy of "solid solution" has not been critically examined, but since Ariya's view has gained considerable qualitative acceptance, it is worth attempting to formulate its terms and implications more precisely, in order to find out whether it is viable as a working hypothesis.

8.2. The microdomain concept: coherent intergrowth

The essence of the microdomain idea is that local order is nearly perfect everywhere, but corresponds to two distinct structures and compositions at different positions within a single ultramicroscopic crystal. Except for the atoms lying in the boundary between two different ordered regions, the environment of every atom is fully ordered. Each region or microdomain has the structure and composition appropriate to a real or idealized compound in the binary system. Thus, in wuestite, microdomains of the perfect B1 structure may be postulated, although this is known only as a high pressure phase (Katsura, Iwasaki and Kimura [1967]). It follows that each microdomain structure could be associated with macroscopic thermodynamic properties. The composition of the nonstoichiometric phase is then determined by the variable proportions in which the two types of microdomain may be present in the dispersion. For convenience, the one structural type may be treated as discrete microdomains in a matrix of the other type, with the proviso that the characteristic dimensions of every region of both microdomains and matrix are very small. Across a wide stoichiometric range the roles of dispersed microdomain structure and matrix structure may well become interchanged.

If the microdomain/matrix assemblage is to behave as monophasic, a necessary condition is that the two structures should be fully coherent. The possibility of coherent intergrowth of two different crystal structures is well known. Wadsley [1964] has discussed the formation of intergrown crystals in which, given sufficient structural and dimensional similarity across some plane of each structure, topotaxy can lead to the formation of microscopic regions of two or more compounds in what is recognizable as a single crystal. At the other extreme, regular repetition of intergrown units of two or more kinds can build up crystals, with large unit cells, which can be regarded as regular intergrowth structures. The composite block structures, such as $W_4Nb_{26}O_{77}$ (Andersson, Mumme and Wadsley [1966]), the bismuthyl perovskite compounds $Bi_2O_2A_{n-1}B_nO_{3n+1}$ (A = K, Na, Ca, Sr,Pb;B = Ti, Nb, Ta) of Aurivillius [1949, 1952] and the remarkable layer ferrites, with c axes up to 1500 Å long (Kohn, Eckart and Cook [1967]) are of this type.

Microdomain coherence must involve small regions, finite in all dimensions, so related that the diffraction symmetry of the whole assemblage is statistically that of a small pseudocell, which represents a common structural framework for both microdomain and matrix. It may be tentatively proposed that a condition for microdomain coherence is that one sublattice is common to and continuous throughout both microdomains and matrix, with only minimal shifts in atomic positions. In the nonstoichiometric fluorite oxides, for example, this would be the cation sublattice; in the B8/C6 chalcogenides, or in $Fe_{1-x}O$, it would be the anion sublattice. The structural differences reside in the other sublattice and heterogeneity at the microdomain level comes from the segregation of point defects within discrete elements of volume, where they are assimilated into a new structural arrangement by ordering, relaxation or non-diffusive reconstruction. It is the diffusively more mobile sublattice which is so modified and, at the interface of each region, units of the structure may be formed or annihilated by unit diffusion processes. In the high temperature dynamic equilibrium the size, shape and position of a given microdomain is subject to fluctuations.

In specific terms, such microdomains of new order could arise in cation deficient structures of the B8/C6 type as a result of the local concentration of unoccupied sites into discs, with the ordering pattern of one of the intermediate phases, or by the collapse of a vacancy disc to form an element of C6 structure, or by formation of an interstitial disc which would constitute an element of pure B8 structure. In the first case, the boundary of the ordered region would approximate to an antiphase boundary; in the second and third cases it would be a dislocation ring. Similarly, in the fluorite structure, anion vacancies may well associate in pairs to generate new $[MO_6]$ octahedra, which aggregate either to create elements of (for example) pyrochlore structure, or to linear strings, as in the lanthanide oxides. Interstitial discs or vacancy discs may involve the surrounding crystal structure in relaxation and displacement to form finite elements of shear planes, bounded by a dislocation ring.

Although these examples are largely conjectural, they show how the observed structural relations in a succession of phases are compatible with the ultramicrostructure proposed by Ariya. They suggest that the necessary condition for microdomain intergrowth can be more precisely defined as requiring perfect coherence on at least two interfaces, the remaining boundaries of each region being either coherent or involving a dislocation ring. Analogy with pre-precipitation phenomena, such as Guinier-Preston zones, suggests that disc-shaped microdomains may well be the energetically favoured shape.

8.3. Thermodynamics of microdomain subdivision

The postulate that microdomains are very small brings them at once within the area of thermodynamics discussed by T. L. Hill [1964], in which the *size* of a system has to be included as one of the extensive variables. In

Hill's terminology, each microdomain is a *small system*, and a crystal or particle of a nonstoichiometric compound is an *ensemble of small systems*. As far as the *subdivision equilibrium* is concerned, it is a closed ensemble; in its equilibrium with the components at a specified chemical potential it is an open ensemble of variable subdivision. An additional degree of freedom follows automatically when some function of size is introduced into the thermodynamic equations. Thus, although each of the two types of small system in the ensemble is invariable in composition, the ensemble as a whole will have bivariant properties.

Standard thermodynamic properties relate to quasi-infinite crystals, for which the discontinuity at the surface affects a negligible proportion of the atoms. In small systems, size enters in the formal guise of a surface free energy term, with some functional dependence on the number of atoms in the system. With the definition of coherence between microdomain and matrix adopted above, it is possible to recognize three components in the "small system" term: (i) A true interfacial energy comes from the discontinuity in the potential series for the electrostatic lattice energy on passing from one structure to another. The change at the interface, from one structure to the other, will give atoms in that interface a proportion of "wrong" next neighbours. The major part of this interfacial energy may thus be related to defects of short range order and site preference energies. (ii) There is a volume strain energy. In the unconstrained state, the two structures will, in general, differ in the dimensions of their common sub-cell. Thus, if UO_{2+x} is described in terms of microdomains of U_4O_9 in a matrix of UO_2, $a_0(UO_2) = 5.470$ Å, $a_0(U_4O_9) = 5.430$ Å. The averaged lattice potential of the nonstoichiometric phase gives a sub-cell with intermediate dimensions, dependent on the composition. In a purely formal sense, the one structure could be regarded as undergoing a homogeneous compression, the other a homogeneous dilation in the microdomain system. The thermodynamic consequences of this can be worked out along the lines laid down by Li, Oriani and Darken [1966]. (iii) If coherence is not complete over the whole interface, an energy is to be associated with the bounding dislocation ring.

8.4. Subdivision equilibrium

The crux of the problem is the *subdivision equilibrium*: the distinction between a microdomain of a new structure and a viable nucleus for crystal growth – i.e. why a microdomain dispersion should be more stable than a macroscopic mixture of two phases. It is particularly important, in this connection, that the thermodynamic stability of nonstoichiometric phases, and their reversible conversion to biphasic mixtures, has been established beyond question (e.g. in the CeO_{2-x} and similar systems).

We consider this question for a system in which, at low temperatures, there are two line phases MX_a (=structure A) and MX_{a+b} (=structure B), and at high temperatures a nonstoichiometric phase MX_n $(a < n < a + b)$. The low temperature equilibrium can be represented as

$$MX_a + \tfrac{1}{2}bX_2 \rightleftharpoons MX_{a+b} . \tag{1.43}$$

A reaction product of total composition MX_n is, under these conditions, a 2-phase mixture, with the phase composition

$$(1 - \xi)MX_a + \xi \, MX_{a+b} \quad \text{where} \quad \xi = (n - a)/b \tag{1.44}$$

and equilibrium is reached for a chemical potential of X_2 represented by $\mu_{X_2}^*$, which is independent of ξ.

At high temperatures, the equilibrium is

$$MX_a + \tfrac{1}{2}(n - a)X_2 \rightleftharpoons MX_n . \tag{1.45}$$

If the total system contains N_M atoms of M, it follows that $\xi \, N_M$ atoms of M are present in dispersed microdomains of structure B, $(1 - \xi)N_M$ atoms of M in the matrix structure A. The equilibrium chemical potential $\mu_{X_2}(\xi)$ is an increasing function of ξ.

The size of each microdomain can be formally defined by the number of lattice molecules (formula units, unit cells etc.) of B structure in it. Let the ensemble contain $\mathcal{N}$ microdomains of mean size $\bar{N}$. In a closed system, with a given number N_M, N_X of component atoms, $\mathcal{N}\bar{N}$ is constant, but both $\mathcal{N}$ and $\bar{N}$ are variable, fluctuating around some average degree of subdivision. If $\bar{N} \to \infty$, $\mathcal{N} \to 1$; the equilibrium state corresponds to the process of nucleation, growth and separation of the two structures as separate phases in the macroscopic sense. If $\bar{N} \to 1$, $\mathcal{N} \to \infty$, the final state is a dispersion of isolated units of structure B, in the form of point defect centres or unit defect clusters.

The thermodynamic condition for a stable disperse state can now be recognized. Classical nucleation theory treats of spontaneous processes – solidification, precipitation, crystallization of a new phase etc. – under non-equilibrium conditions, in which the forward reaction produces a nett decrease in free energy. For equilibrium in equation (1.43) $\Sigma v_i \mu_i = 0$, where the chemical potentials assigned to compounds A and B relate to the substances in their standard states. For the formation of very small crystallites or embryos of B $\Delta G_B > \Delta G_B^0 = \Delta G_B^0 + \Delta G_B^{ex}$, where the excess free energy represents the size-dependent interfacial free energy: $\Delta G_B^{ex} = f(N)$. If reaction (1.43) were carried out isothermally and reversibly in the thermodynamic sense, with $\mu_{X_2} \equiv \mu_{X_2}^*$, $\Delta G > 0$ for all nucleus sizes; the size of the embryos would then conform to a Gibbs distribution. If $\mu_{X_2} > \mu_{X_2}^*$ there will be some critical nucleus size, above which $\Sigma v_i \mu_i < 0$, so that reaction, nucleation and crystal growth proceed spontaneously. The condition that a disperse state of small microdomains should be stable (eq. (1.45)) is that some property of the system leads to equilibrium at $\mu_{X_2}(\xi) < \mu_{X_2}^*$.

Accepting the microdomain idea in no other sense than as a working hypothesis, a negative excess chemical potential $\mu_{X_2}^{ex}(\xi) = \mu_{X_2}(\xi) - \mu_{X_2}^*$ may be assigned to the mobile component in the nonstoichiometric equilibrium (1.45). It is then possible to examine the supposed subdivision equilibrium before enquiring the origin and magnitude of what emerges as a necessary condition that the postulate should be valid.

The subdivision equilibrium can be treated by the formalism developed by Hill for small system thermodynamics, but the same result is reached more clearly by the method used in Frenkel's discussion of nucleation. Let m formula units of structure A be required for the formation of 1 lattice molecule of structure B. (E.g., if the Willis cluster, involving one complete unit cell of UO_2, is taken as the unit structure of microdomains in UO_{2+x}, $m = 4$.) Then the formation of microdomains of size N could be represented as requiring (a) the establishment of a very dilute point defect equilibrium between the components (1.46) and (b) reaction and aggregation to form the N-fold microdomains (1.47).

$$A + \tfrac{1}{2}X_2 \rightleftharpoons X_I \text{ (in A)} \quad \text{chemical potentials } \mu_{X_2}, \mu_A^0 \tag{1.46}$$

$$mNA + mbNX_I \rightleftharpoons (B)_N \quad \text{aggregation,} \tag{1.47}$$

or

$$mNA + \tfrac{1}{2}mbNX_2 \rightleftharpoons (B)_N \quad \text{nett reaction.} \tag{1.48}$$

The equilibrium size distribution amongst microdomains of B can be represented as reached by a set of unit step reactions (1.49):

$$(B)_N \quad \rightleftharpoons (B)_{N-1} + mA \quad + \tfrac{1}{2}mb X_2 \tag{1.49}$$
$$(B)_{N-1} \rightleftharpoons (B)_{N-2} + mA \quad + \tfrac{1}{2}mb X_2$$

$$\begin{array}{cccc} \cdot & \cdot & \cdot & \cdot \\ \cdot & \cdot & \cdot & \cdot \\ \cdot & \cdot & \cdot & \cdot \end{array}$$

$$(B)_{N_0+1} \rightleftharpoons (B)_{N_0} + mA \quad \tfrac{1}{2}mb X_2$$
$$(B)_{N_0} \quad \rightleftharpoons \quad mN_0 A + \tfrac{1}{2}mb N_0 X_2$$

where the smallest possible size of microdomain, as distinct from a point defect, is N_0 lattice molecules. If the lattice molecules of the B structure are the defect clusters, N_0 may be 1.

The free energy changes in the series of subdivision reactions (1.49) can be summed if the (admittedly questionable) approximation is made that each microdomain of size N can be treated as a solute species, in the matrix as solvent, with an ideal entropy of solution. The mole fraction x_N in each size group can then be evaluated as

$$x_N = x_A^N \exp\left[\frac{Nmb\mu^{ex}}{kT}\right] \exp\left[-\frac{\sigma f(N)}{kT}\right] \tag{1.50}$$

where x_A is the mole fraction of solvent (i.e. matrix structure A) and $\sigma f(N)$ includes all contributions to the small system or interfacial energy of the N-sized microdomains.

Writing $\delta = \mu^{ex}/kT$, it can be shown that (1.50) leads to an average microdomain size $\bar{N}$ as in (1.51):

$$\bar{N} - N_0 + 1 = \frac{1}{1 - x_A(1 - mb\delta)} \exp\left[-\frac{\sigma f(N)}{kT}\right]. \tag{1.51}$$

It is impossible to deal explicitly with the last term in (1.51) without some specific hypotheses about the effective interfacial energy, shape of microdomains etc. However, the optimum condition for the formation of large microdomains should be perfect, strain-free coherence – i.e. $\sigma \to 0$. If this limiting assumption is made,

$$\bar{N} - N_0 + 1 = 1/\{1 - x_A(1 - mb\delta)\}. \tag{1.52}$$

From a mass balance, x_A can be expressed in terms of ξ and $\bar{N}$:

$$x_A = \frac{m\bar{N}(1 - \xi)}{m\bar{N}(1 - \xi) + \xi}. \tag{1.53}$$

$\bar{N}$ can then be evaluated as a function of μ^{ex}/kT for any value of ξ.

Results of such a calculation (Fig. 1.24) make it clear that, unless μ^{ex}/kT is very small indeed, the average size of microdomains is very small and strongly biased in favour of the smallest units. It is then possible to

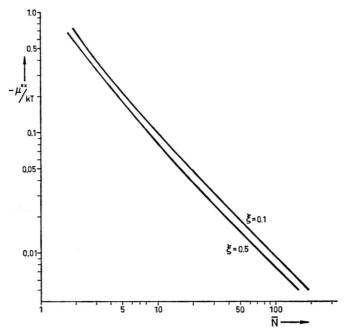

Fig. 1.24. Variation of average microdomain size $6/N$ with μ^{ex}/kT.

enquire how this conclusion relates to real systems. Fig. 1.25a shows a schematic T, X projection of phase equilibria in the type of system considered. As long as phases A and B are constant in composition,

$$\mu^*_{X_2}(T) = 2/b\{\mu^0_B(T) - \mu^0_A(T)\} \qquad (1.54)$$

and $\mu^*_{X_2}$ is a nearly linear function of the temperature (Fig. 1.25b), independent of the total composition MX_n, up to that temperature T_1 at which that composition enters the nonstoichiometric range and becomes monophasic. Above T_1, $\mu_{X_2}(n, T)$ falls on a new line, of different slope. Then, at any temperature, $\mu^* - \mu(n)$ represents the decrease in μ_{X_2} on transforming a

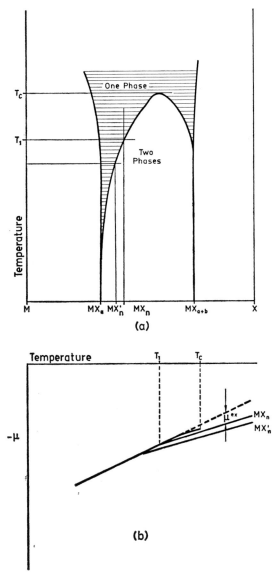

Fig. 1.25. (a) $T–X$ equilibrium diagram for system with nonstoichiometric compounds and critical miscibility above T_c; (b) Corresponding $\mu–T$ diagram, showing divergence of $\mu(\xi,T)$ from $\mu^*(T)$, the composition-independent chemical potential in 2-phase mixtures of AB and AB$_2$.

References p. 74

biphasic mixture of $A + B$ into the single nonstoichiometric phase. $\mu^{ex}(T)$ thus necessarily increases rapidly (and linearly) with temperature, from the value zero at T_1.

8.5.

A severe limitation of this treatment is that the observed μ^{ex} is equated entirely to the contribution made by the entropy of solution of microdomains in the matrix and this is probably unjustifiable. In any case, comparison with real systems meets the difficulty that (as for CeO_{2-x} and UO_{2+x}) the equilibria in the biphasic range have usually been measured only at high temperatures, at which the coexisting phases A', B' already deviate significantly from ideal composition. The μ^*–T curve is then neither strictly linear nor given by $\mu_A^0 - \mu_B^0$.

However, μ^{ex}/kT is certainly very small for compositions and temperatures close to the 2-phase envelope. The supposition made by Gerdanian and Dode ([1965], Kotlar, Gerdanian and Dode [1967]) to explain the shape of their $(p,X)_T$ isotherms for the UO_{2+x} system, close to the U_4O_9–UO_{2+x} coexistence conditions, can therefore be justified. However, at temperatures further removed from phase coexistence, the mean size of microdomains would appear to decrease rapidly. Taking the data for the UO_{2+x} system, it would appear that, at temperatures 100° above the solvus line, for any composition $UO_{2.05}$ to $UO_{2.15}$, $-\mu^{ex}/kT > 1$. This would indicate that $\overline{N} < 2$, with the average decreasing yet further at higher temperatures. If this is significant, it would imply that the Willis clusters behave as independent defects and are not appreciably ordered into blocks in the high temperature phase. For the CeO_{2-x} system, using the data of Bevan and Kordis, the change in the slope of the $\mu(T)$ line, in passing from univariant to bivariant conditions, is much smaller. However, the $\mu_{O_2}^*(T)$–T line is defined with less certainty, because of the lack of experimental measurements below the α-phase range. It can be roughly deduced that, for $CeO_{1.94}$, microdomain sizes would be:

T	920°	1000°	1050°K
μ^{ex}/kT	0.15	0.20	0.25
$\overline{N}$	7	5.5	4.5

These figures may represent some aggregation of defect centres in the α-phase, but are again too small to be indicative of Ariya's "islands of new structure". The aggregates might better be described as extended clusters ("multiclusters"), or perhaps as the fragmentary strings of $[MO_6]$ octahedra discussed by Hyde, Bevan and Eyring.

9. Conclusion

Although average microdomain sizes have been inferred by a rather arbitrary treatment of the data, it seems inescapable that any ordered elements of structure in nonstoichiometric phases must be on a much smaller scale than was proposed by Ariya (200–1000 unit cells in the "island" block). However, in reducing the qualitative ideas to a form suitable for critical examination, some over-restrictive conditions have been imposed. In particular, the earlier discussion of fluctuations in ordered phases (p. 23) suggests that a multiplicity of ordered states should be considered. More serious is the problem that only the subdivision equilibrium can be analysed in purely thermodynamic terms; two main characteristics of nonstoichiometric compounds undoubtedly require extra-thermodynamic interpretation – specific structural and statistical mechanical models. The first is the saturation of matrix phase with microdomains – i.e. the limited accessible composition range below the critical temperature. The foregoing treatment was in terms of ideal solution theory; the observed behaviour is certainly non-ideal. The second problem is that, within an ideal solution theory, microdomain size has been related to an excess chemical potential, without further explanation. It is possible that a more specific model for a non-ideal microdomain system could be framed which would remove the arbitrary elements and, perhaps, lead to an upward revision of the estimates for the size of ordered elements of structure.

Nevertheless, it is of interest that the microdomain model and the treatments of random, interacting defects appear to converge in their representation of nonstoichiometric phases. The need for a satisfactory theoretical model, to correlate the observed structural and thermodynamic properties remains – even though (as will be evident from earlier sections) there are difficulties in stringently testing the validity of alternative hypothesis. It may well be that the direction most amenable to theoretical development is still the treatment of ordering within an initially randomized system, rather than the hypothesis of perfect local order.

References p. 74

References

ACKERMANN, R. J. and R. W. SANDFORD, 1966, Argonne National Laboratory Report ANL-7250.

ALLNATT, A. R. and M. H. COHEN, 1964, J. Chem. Phys. **40**, 1860, 1871.

ALLPRESS, J. G., J. V. SANDERS and A. D. WADSLEY, 1969, Acta Cryst. **B25**, 1156.

ALYAMOVSKII, S. I., P. V. GEL'D, G. P. SHVEIKIN and E. N. SHCHETNIKOV, 1968, Russ. J. Inorg. Chem. **13**, 472.

ANDERSON, J. S., 1946, Proc. Roy. Soc. **A185**, 69.

ANDERSON, J. S., 1964, Proc. Chem. Soc. 166.

ANDERSON, J. S. and B. G. HYDE, 1967, Phys. Chem. Solids **28**, 1393.

ANDERSON, J. S. and A. S. KHAN, 1969, unpublished.

ANDERSSON, S., W. G. MUMME and A. D. WADSLEY, 1966, Acta Cryst. **21**, 802.

ANDERSSON, S., 1967, Arkiv Kemi **26**, 521.

ARIYA, S. M. and M. P. MOROZOVA, 1958, Russ. J. Gen. Chem. **28**, 2647.

ARIYA, S. M. and Y. G. POPOV, 1962, Russ. J. Gen. Chem. **32**, 2077.

ATLAS, L. M., 1968a, Phys. Chem. Solids **29**, 91.

ATLAS, L. M., 1968b, Phys. Chem. Solids **29**, 1349.

ATOJI, M. and M. KIKUCHI, 1968, Argonne National Laboratory Report ANL-7441.

AURIVILLIUS, B., 1949, Arkiv Kemi **1**, 463, 499.

AURIVILLIUS, B., 1952, Arkiv Kemi **5**, 39.

BERTAUT, E. F., 1953, Acta Cryst. **6**, 557.

BEVAN, D. J. M., 1955, J. Inorg. Nucl. Chem. **1**, 49.

BEVAN, D. J. M. and J. KORDIS, 1964, J. Inorg. Nucl. Chem. **26**, 1509.

BEVAN, D. J. M., R. S. CAMERON, A. W. MANN, G. BRAUER and U. ROETHER, 1968, Inorg. Nucl. Chem. Letters **4**, 241.

BILZ, H., 1958, Z. Physik **153**, 338.

BØHM, F., F. GRØNVOLD, H. HARALDSEN and H. PRYDZ, 1955, Acta Chem. Scand. **9** , 1510.

BORGHESE, C., 1967, Phys. Chem. Solids **28**, 2225.

BRAUER, G. and K. A. GINGERICH, 1957, Angew. Chemie **69**, 480.

BRAUER, G., K. A. GINGERICH and U. HOLTSCHMIDT, 1960, J. Inorg. Nucl. Chem. **16**, 27.

BRAUER, G. and K. A. GINGERICH, 1960, J. Inorg. Nucl. Chem. **16**, 87.

BREBRICK, R. F., 1958, Phys. Chem. Solids **4**, 190.

BREBRICK, R. F., 1959, Phys. Chem. Solids **11**, 43.

BREBRICK, R. F., 1961, Phys. Chem. Solids **18**, 116.

BREBRICK, R. F., 1963, Phys. Chem. Solids **24**, 27.

BREBRICK, R. F., 1967, Non-Stoichiometry in Binary Semiconductor Compounds, $M_{\frac{1}{2}-\delta}N_{\frac{1}{2}+\delta}$ (c), in: Progress in Solid State Chemistry, Vol. 3, ed. H. Reiss (Pergamon Press) pp. 213–264.

BURCH, R., 1969, unpublished calculations.

CAHN, J. W., 1961, Acta Met. **9**, 795.

CLARK, A. H., 1965, Nature **205**, 792.

DENKER, S. P., 1964, Phys. Chem. Solids **25**, 1397.

DENKER, S. P., 1968, J. Less Common Met. **14**, 1.

EHRLICH, P., 1949, Z. Anorg. Allgem. Chem. **260**, 19.

ELCOCK, E. W., 1956, Order Disorder Phenomena (Methuen, London).

ELLIOTT, G. R. B. and J. F. LEMONS, 1963, Advan. in Chem. **39**, 144, 153.

EROFEEVA, M. S., N. L. LUKINYKH, and S. M. ARIYA, 1961, Russ. J. Phys. Chem. **35**, 375.

GERDANIAN, P. and M. DODE, 1965, J. Chim. Phys. **62**, 171.

GOLDSCHMIDT, H. J., 1967, Interstitial Alloys (Butterworth's, London).

GRØNVOLD, F., H. HARALDSEN and J. VIHOVDE. 1954, Acta Chem. Scand. **8**, 1927.

GRØNVOLD, F. and E. JACOBSEN, 1956, Acta Chem. Scand. **10**, 1440.

GRØNVOLD, F. and E. F. WESTRUM, 1959, Acta Chem. Scand. **13**, 241.

GRØNVOLD, F., 1968, Acta Chem. Scand. **22**, 1219.

HAGEMARK, K. and M. BROLI, 1966, J. Inorg. Nucl. Chem. **28**, 2837.

HERAI, T., B. THOMAS, J. MANENC and J. BENARD, 1964, Compt. Rend. **258**, 4528.

HILL, T. L., 1963, Thermodynamics of Small Systems (Benjamin, N.Y.).

HOCH, M., A. S. IYER and J. NELKEN, 1962, Phys. Chem. Solids **23**, 1463.

HOCH, M., 1963, Phys. Chem. Solids **24**, 157.

HYDE, B. G., D. J. M. BEVAN and L. EYRING, 1966, Phil. Trans. Roy. Soc. **A259**, 583.

JELLINEK, F., 1957, Acta Cryst. **10**, 620.

KATSURA, T., B. IWASAKI and S. KIMURA, 1967, J. Chem. Phys. **47**, 4559.

KEDROVSKII, O. V., I. V. KOVTUNENKO, E. V. KISELEVA and A. A. BUNDEL', 1967, Russ. J. Phys. Chem. **41**, 205.

KOHN, J. A., D. W. ECKART and C. F. COOK, 1967, Mat. Res. Bull. **2**, 55.

KOTLAR, A., P. GERDANIAN and M. DODE, 1967, J. Chim. Phys. **64**, 862.

KOZLENKO, T. A., P. V. KOVTUNENKO, E. V. KISELEVA and A. A. BUNDEL', 1967, Russ. J. Phys. Chem. **41**, 588.

KRÖGER, F. A., 1964, The Chemistry of Imperfect Crystals (North-Holland Publ. Co, Amsterdam).

KRÖGER, F. A., 1968, Phys. Chem. Solids **29**, 1889.

LI, J. C. M., R. A. ORIANI and L. S. DARKEN, 1966, Z. Physik. Chem. **NF49**, 271.

LIBOWITZ, G. G., 1962, J. Appl. Phys. Suppl. **33**, 399.

LIBOWITZ, G. G., 1963, A. C. S. Advances in Chemistry **39**, 75.

LIDIARD, A. B., 1962, J. Appl. Phys. Suppl. **33**, 414.

LOTGERING, F. K., 1955, Z. Physik. Chem. **NF4**, 238.

MANENC, J., G. VAGNARD and J. BENARD, 1962, Compt. Rend. **254**, 1777.

MARTIN, S. L. H. and A. L. G. REES, 1954, Trans. Far. Soc. **50**, 343.

MESSER, C. E. and G. WAN-HOI HUNG, 1968, J. Phys. Chem. **72**, 3958.

OKAZAKI, A., 1959, J. Phys. Soc. Japan **14**, 112.

OKAZAKI, A., 1961, J. Phys. Soc. Japan **16**, 1162.

REES, A. L. G., 1954, Trans. Far. Soc. **50**, 335.

ROBERTS, L. E. J. and A. WALTER, 1961, J. Inorg. Nucl. Chem. **22**, 213.

ROOF, R. B. and G. R. B. ELLIOTT, 1965, Inorg. Chem. **4**, 691.

ROTH, R. S. and A. D. WADSLEY, 1965, Acta Cryst. **18**, 724.

ROTH, R. S. and A. D. WADSLEY, 1965, Acta Cryst. **19**, 26.

SCHREIBER, D. S. and R. M. COTTS, 1963, Phys. Rev. **131**, 1118.

SPEISER, R. and J. W. SPRETNAK, 1955, Trans. Am. Soc. Metals **47**, 493.

STRIPP, K. F. and J. G. KIRKWOOD, 1954, J. Chem. Phys. **22**, 1579.

THORN, R. J. and G. H. WINSLOW, 1966, J. Chem. Phys. **44**, 2632.

WADSLEY, A. D., 1964, Non-Stoichiometric Compounds (Academic Press, N.Y.) p. 98.
WADSLEY, A. D., 1967, Helv. Chim. Acta, Fasc. extraord. 207.
WATANABE, D., J. R. CASTLES, A. JOSTSENS and A. S. MALIN, 1967, Acta Cryst. **23**, 307.
WESTMAN, S. and C. NORDMARK, 1960, Acta Chem. Scand. **14**, 465.
WESTMAN, S., 1962, U. S. Dept. Army Tech. Rept. DA-91-591-EUV-1319.
WOJCIECHOWSKI, K. F., 1961, Physica **27**, 509.

CHAPTER 2

RECENT PROGRESS IN THE INVESTIGATION
OF NONSTOICHIOMETRY

Y. P. JEANNIN

Département de Chimie Inorganique, Faculté des Sciences de Toulouse, Toulouse, France

1. Introduction

To observe that the chemical formula of a crystallized compound does not follow the law of definite proportions proposed by Proust, and that the ratio of atom numbers defined by this formula may change continuously while the unit cell seems to stay the same apart from slight variations in parameters, is now very common. This field of research has rapidly grown and has been developed to a large extent since the publication of the well known paper by Wagner and Schottky [1931], in which various kinds of elementary punctual lattice defects, vacancy, substitution, and additional inserted atom, have been postulated and described and since experimental techniques allowed of accurate and sensitive study of the solid state. Thus the idea of a strictly definite stoichiometric compound has lost part of its value, because today such a compound appears as one particular composition of homogeneous phase, or even sometimes as an asymptotic limit.

The importance of deviations from stoichiometry, that is the extent of the homogeneous field, is very variable from one system to another; in fact it may be so narrow that it escapes chemical detection and is only noticeable with physical methods; but it may be so wide that interactions between defects become important and that short range ordering, sometimes involving a reconstruction of the lattice has to be considered; as a limit, ordering may be achieved.

Systems among which broad nonstoichiometric phases are frequently met are those constituted by transition metals with nonmetallic elements of the sixth column of the periodic table.

The different problems which arise when point defects are introduced in

References p. 126

a lattice, will be considered. Examples will be mainly chosen among binary compounds. As a matter of fact, they have given rise to a large number of research projects in the last few years. Several extensive reviews have recently appeared, emphasizing either a classifying point of view (Wadsley [1964]), or a particular point of view, such as compounds related to the nickel arsenide type (Kjekshus and Pearson [1964]), semiconducting compounds (Brebrick [1967]), or new results (Libowitz [1965]). The aim of this chapter is to present a comprehensive, rather than an extensive survey of the recent results which will allow an analysis of some of the new advances and ideas brought to the problem of nonstoichiometry. In the first part, we shall introduce problems related to the existence of homogeneous phases. In them, the number of defects may become so large that they interact; as a result, ordering phenomena which produce superstructures may occur: this will be the subject of the second part.

2. Homogeneous phases

2.1. Thermodynamic point of view

The introduction of defects modifies the free enthalpy in two ways. Firstly, changing enthalpy, as for instance drawing an atom out of its normal lattice site is an endothermal process. Secondly, increasing entropy as there are several possible distributions of defects, assumed spread out statistically. At a given temperature, if the number of defects grows, the energy expended in creating them increases; at the same time, the configurational entropy of the system increases. Variations of enthalpy and entropy, continuous functions of the number of defects act in opposite directions and make the value of free enthalpy $G = H - TS$ minimum for a given value of the number of defects. Consequently, an ideal and perfect lattice does not represent, thermodynamically speaking, the most stable energetic situation at a nonzero temperature.

On the other hand, it is important to note that any $A_m B_n$ solid compound belongs to a nonstoichiometric field, due to a continuous change in the atomic ratio n/m because of point defects, if temperature is not absolute zero. This can be shown representing the curves of free enthalpies of liquid G_{liq} and solid G_{sol} as a function of composition at given temperature and pressure (Fig. 2.1). Each curve presents a minimum related to a maximum stability of the abscissa composition. Change of free enthalpy G_{sol} with composition is more or less rapid, depending on the compound; in other words, minimum is more or less sharp. If both curves G_{liq} and G_{sol} are plotted in such a way that the curve related to the solid goes below that of the liquid, as in Fig. 2.1, the solid must crystallize. Common tangents to

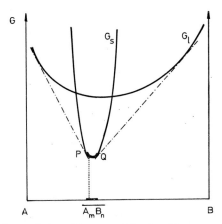

Fig. 2.1. Variation of free enthalpy with composition in a binary system. G_l is related to a liquid mixture of A and B. G_s is related to a solid of variable composition. PQ-segment represents the stability range of a nonstoichiometric compound; the minimum of G_s does not correspond to the stoichiometric composition A_mB_n.

G_{liq} and G_{sol} curves can be drawn; these lines show the identity of chemical potentials in both liquid and solid phases. Between tangency points of the G_{sol} curve, there is a single-phased range of stability for a solid of variable composition. Understood in such a way, any solid binary compound shows deviation from stoichiometry. Obviously, the same discussion might be developed about a ternary compound in three-dimensional space. The width of the homogeneous phase is related firstly to the respective positions of the two curves G_{liq} and G_{sol}, that is to temperature, and secondly to the more or less wide aperture of the G_{sol} curve, that is to the nature of the chemical bond in A_mB_n and to the nature of defects in solid A_mB_n.

The variation in composition of A_mB_n is possible through three elementary kinds of point defects; vacancies on A or B lattice changing the formula in $A_{m-x}B_n$ or A_mB_{n-x}; additional inserted atoms A or B giving $A_{m+x}B_n$ or A_mB_{n+x}; antistructural defects, that is surplus atoms A in place of atoms B, or surplus atoms B in place of atoms A, giving $A_{m+x}B_{n-x}$ or $A_{m-x}B_{n+x}$

References p. 126

(Fig. 2.2). For instance, nonstoichiometric FeO (iron II oxide) contains iron vacancies (Jette and Foote [1933]; Hägg [1933]). Nonstoichiometric TiS_2 (titanium disulfide) contains additional titanium atoms (Jeannin and Bénard [1959]). Nonstoichiometric zinc antimonide, ZnSb, shows antistructural defects through an excess of zinc (Bokii and Klebtsova [1965]).

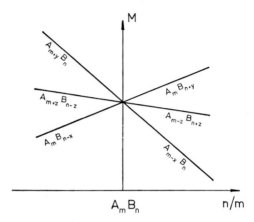

Fig. 2.2. Variation of the mass contained in the unit cell when composition changes around A_mB_n for different defect types: vacancy, interstitial, substitution.

From free enthalpy curves, it is clear that there is no reason that the stoichiometric composition necessarily corresponds to the minimum of G_{sol} curve or, in other words, that the stoichiometric composition is the most stable; this to the same remark as that previously made about the fact that minimum free enthalpy appears when some defects occur in the lattice. Even more, the stoichiometric composition may be outside the homogeneous phase, which means that it does not exist in this case. For instance, strictly stoichiometric ferrous oxide appears not possible to prepare (Aubry and Marion [1955]; Engell [1957]) except under high pressure (Katsura et al. [1967]). Cuprous sulphide is probably not stable, although one can prepare sulfides having a composition very close to stoichiometry such as $Cu_{1.9996}S$ (Wagner and Lorenz [1957]; Wagner and Wagner [1957]). Besides, the maximum melting point does not correspond to stoichiometry if both liquidus and solidus curves of free enthalpy as function of composition are not symmetrical with respect to stoichiometry, and indeed they have no reason to be so systematically. Tin telluride which has the highest melting point contains 50.4 tellurium atoms per 100 atoms (Brebick [1963]; Brebick and Strauss [1964]).

In order to give a complete description of a nonstoichiometric field, it is necessary to determine the extent of the field and then to find the nature of defects responsible for deviation from stoichiometry. Thermodynamic

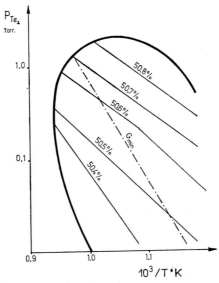

Fig. 2.3. Range of stability of nonstoichiometric tin telluride in pressure–temperature plane. Values are related to the tellurium content in atom per cent. G_{min} never corresponds to stoichiometry. (By courtesy of Brebick and Strauss and of the American Institute of Physics.)

measurements, as well as other physico-chemical methods, or better still, a combination of these methods, when necessary, may be used for that purpose.

Particularly, with the help of thermodynamic measurements, the width of the homogeneous field of tin telluride has been determined by Brebick and Strauss [1964]. The vapour pressure of tellurium, in equilibrium with the solid, is plotted against the reciprocal of absolute temperature. This diagram (Fig. 2.3) shows the region of the (p,T) plane where nonstoichiometric tin telluride is stable. Let us point out that the minimum free enthalpy curve never corresponds to stoichiometry.

More can be deduced from these thermodynamic measurements, for instance, the determination of the enthalpy of the creation of a point defect

References p. 126

by carefully discussing experimental data. Thus cobalt sulphide, CoS, exhibits deviation from stoichiometry which can be described through the experimental study of the equilibrium phase CoS-limit which is rich in sulphur of Co_9S_8 phase-H_2-H_2S (Laffitte [1959]). It has to be pointed out that Co_9S_8 presents negligible deviation from stoichiometry, which means that this sulfide may be regarded as being stoichiometric. Applying the principle of the initial and final state this permits of the determination of the formation energy of a cobalt vacancy V_{Co} which corresponds to the equation:

$$\tfrac{1}{2}S_2 \leftrightharpoons CoS + 2e^+ + V_{Co}'' .$$

This enthalpy is equal to 0.73 eV.

The case of Digenite, that is the high-temperature allotropic form of Cu_2S, is quite interesting because Rau [1967] has been able to push the description further: not only has the formation enthalpy of defects been determined, but associated defects have also been introduced to explain completely experimental results and their concentration has been given as a function of the deviation from stoichiometry. The experimental dependance of the logarithm of sulphur vapour pressure with the sulphur content of the solid in equilibrium with a vapour phase, either sulphur or hydrogen sulfide, has been determined and a binomial formula fitted to that curve. Then, applying the mass action law to two equilibria, the first corresponding to the creation energy of an effectively neutral copper vacancy

$$\tfrac{1}{4}S_2 \leftrightharpoons V_{Cu}^* + \tfrac{1}{2}Cu_2S$$

and the second related to the ionization of that vacancy liberating a positive hole

$$V_{Cu}^* \leftrightharpoons V_{Cu}' + e^+ .$$

It is found that the same binomial formula has been experimentally derived by Rau. Looking at these two experimental and theoretical formulae together, it becomes possible to determine firstly the creation energy of an effectively neutral copper vacancy, equal to $-5365 + 3.60T$ in calories per mole that is 0.08 eV at 700°C, and secondly the ionization energy of this neutral vacancy equal to 0.55 eV. Moreover, a careful examination of curves shows that there is some discrepancy between experiments and the model described above, particularly at high temperature and for strong deviation from stoichiometry. Rau attributed this to the appearance of a new type of imperfection in appreciable concentration.

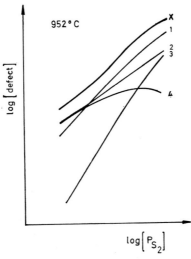

Fig. 2.4. Logarithm of defect concentration in digenite as a function of the logarithm of sulphur vapour pressure. $\times$ is related to the total number of defects, 1 to the neutral copper vacancies, 2 to the positive holes, 3 to the negatively charged copper vacancies and 4 to a complex defect made of a negatively charged copper vacancy associated with a copper interstitial atom. (By courtesy of Rau and of Pergamon Press.)

This defect may be described as follows: strong polarization induced by a Cu^{2+} ion loosens a close Cu^+ ion, which may then leave the lattice more easily. This yields a complex defect consisting of the association of a neutral and a negatively charged copper vacancy. Thus high electronic conduction of digenite (Wehefritz [1960]; Kamigaichi [1952]) is explained, because such an association is electrically compensated by positive holes. Again mass action law may be applied to this complex situation and concentrations of different types of defects, neutral copper vacancies, ionized copper vacancies, positive holes, complex associated defects, can be computed: they are shown in Fig. 2.4, varying with composition.

Sometimes the simultaneous use of two methods, for instance thermo-dynamical and electrical measurements, brings valuable results. This is particularly applied to semiconducting materials. For example, a diagram showing respective concentrations of different types of defects has been established by Fujimoto and Sato [1966] for lead telluride (Fig. 2.5). Ionized-vacancy-creation energy is 0.3 eV for lead and 1.2 eV for tellurium. It is worth noting that the necessary energy for a metallic atom is much less than

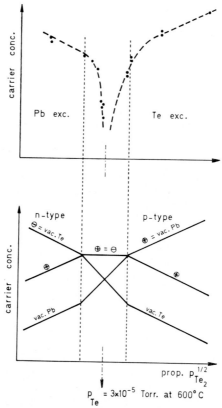

Fig. 2.5. In the upper part the variation of electrical conductivity of PbTe, shown as carrier concentration, is a function of the number of defects related to tellurium vapour pressure. In the lower part are plotted concentrations of different types of defects, electrons, positive holes, lead and tellurium vacancies. (By courtesy of Fujimoto and Sato and of Japanese Journal of Applied Physics.)

that generally required for the nonmetallic atom in this type of compound (Brebrick [1967]).

Similarly, the defect type of cobalt monoxide has been studied by means of electrical conductivity methods under variable oxygen pressure (Fisher and Tannhauser [1966]). These authors have been able to determine the energy necessary to create a singly-ionized vacancy: it is equal to 0.56 eV roughly divided into 0.1 eV, creation energy of a neutral vacancy, and 0.45 eV, its energy of first ionization. The energy of second ionization is given as 0.65 eV.

The following step would be to express thermodynamical functions of the nonstoichiometric solid as a function of the number of defects. It might then be thought that the number and perhaps also the nature of defects could be learnt. A theoretical model would have to be designed and thermodynamic functions computed which would then be compared with experimental data. The problem is not simple and needs methods of statistical thermodynamics for application. Since the original treatment of Wagner and Schottky [1931], several tests have been carried out for different special cases.

Recently Libowitz and Lightstone [1967] tried, with only the knowledge of the activity of one constituent of a binary compound MX_n, to describe a method giving the nature of defects. Mathematical formulae giving the value of activity have been obtained following two different approaches. Firstly, the value of the total free enthalpy of the crystal is written as a function of the free enthalpies of M and X atoms in the perfect crystal and of the number of all possible kinds of defects, vacancies, substitutions, insertions of both atomic species M and X. In writing these equations, the assumption is made that there are no interactions between defects, so that configurational entropy is the same as for ideal mixing. Then, considering successively that each defect is predominant with respect to the others and, applying valuable approximations, the activity of one chemical species, which can be obtained experimentally, may be computed. Secondly, the equations, derived in advance for activities, have been calculated more simply from a mass-action treatment, writing pseudochemical equilibria related to the mechanism describing the formation of each type of defect.

Corresponding equations so derived are quite identical, and have been consistently applied by these authors to experimental data. The experimental curve is drawn at the same time as three theoretical curves, each corresponding to a predominant defect type, vacancies, substitutions, interstitials. Several examples are shown by Libowitz and Lightstone, not all of them being conclusive unfortunately. Indeed, each curve activity as function of deviation from stoichiometry for a given defect type has its own slope. The scattering of experimental data and the relative close slopes of some computed curves also limit the ability to derive the nature of defects straight-

forwardly. For instance (Fig. 2.6), if it is clear that uranium hydride (Libowitz and Gibb [1957]) contains hydrogen vacancies, the situation cannot be
described in that way for $CeCd_{6+x}$ alloy (Elliott and Lemons [1960]). It has
to be pointed out that electronic charge effects were completely neglected;

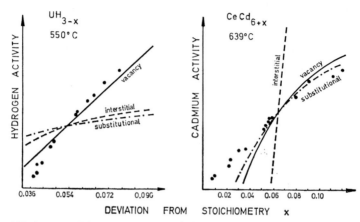

Fig. 2.6. Hydrogen activity in UH_{3-x} and cadmium activity in $CeCd_{6+x}$ shown as a function
of the deviation from stoichiometry x. Points are experimental data. Curves are related to
theoretical computations. (By courtesy of Libowitz and Lightstone and of Pergamon
Press.)

thus, these results cannot be applied to compounds having a high degree of
ionicity or a semiconducting behaviour.

The preceding treatment does not include defect interaction. For large
deviation from stoichiometry, it may not then be correct. To take account of
defect interaction is a difficult problem. The mode of interaction may be
multiple and a general model is difficult to translate into mathematical terms.
The tests by Anderson [1946] and Rees [1954], very well known, have led to
successful results because starting assumptions fit the specially investigated
problem particularly well. One cannot write that, at present, theoretical
relations have been derived which are reliable for any experimental case.

Let us mention a paper by Atlas [1968] who uses a statistical approach.
This treatment is particularly interesting because the proposed model takes
partial ordering into account. For the sake of simplicity oxygen–deficient
metal dioxides of the fluorite type have been chosen, and only oxygen
deviation has been considered. The generalized partition function of defects
takes into account two repulsion energies, that between vacancies and

between reduced cations formed because electrons dissociated from vacancies are localized at cation sites. Defect interaction is introduced in view of the fact that the net energy of a vacancy depends firstly on the number of other defects, vacancies and reduced cations enclosed in a given volume centred on one vacancy, secondly on distances between that vacancy and surrounding defects, and thirdly on the relative charge of reduced cations and of the vacancy. This treatment is therefore purely based upon electrostatic energy considerations.

A local ordering brings into a given volume an excess of defects, vacancies and reduced cations and among them interactions occur. The corresponding energy varies from zero to an appreciable value, simultaneously with defect concentration. It has been set by Atlas that this range of energy may be divided into discrete levels, each having its own arrangement independent of any other level, that is its own temperature dependent population. On this basis, a generalized partition function of defects has been calculated. From this, relative partial molar thermodynamic functions of oxygen in MeO_{2-x} may be computed. Calculated and experimental curves of entropy and free enthalpy functions are shown in Fig. 1.20. When deviation from stoichiometry is less than 0.2, agreement is fairly good and encouraging. Although divergence occurs above this value, there is still qualitative interest because reversing or steepness of slope of computed curves coincides with the existence of a two-phase region in the cerium–oxygen diagram.

Atlas [1969], pursuing his efforts, has tried to correlate the two defect populations, considering ionic compounds and particularly uranium dioxide as example. Distribution functions for two assemblies of defects have been derived. The interaction energy of defects has been estimated on a coulombic way, using an adaptation of Evjen's computation of Madelung constant; an empirical less important term has been added to take account of strain and of overlap repulsion. Hence the model describes the crystal as consisting of small 2:1:2 clusters containing excess anions, anion vacancies paired with displaced anions, and trapped positive holes. The experimental curves describing the change of relative partial molar entropy and enthalpy of oxygen in UO_{2+x} appear in quite good agreement with curves computed. To describe partial ordering in other systems with this special statistics thus appears as a promising possibility.

Thus, thermodynamic considerations, either classical or statistical, are very helpful in the study of nonstoichiometry. Starting from experimental measurements, such as dissociation pressure or electromotive cells, not only can the extent of a nonstoichiometric phase be determined, but defect-creation energy also and sometimes the nature of defects recognized. The problem of association of defects, although difficult, is likely to be attacked along those lines.

2.2. Remarks on preparation of nonstoichiometric compounds

The study of deviation from stoichiometry implies that any experimental measurement has to be made on a chemical compound set in a state of thermodynamic equilibrium. Nonstoichiometric compounds have first to be synthetized, which brings one or more solid phases into reaction. On the other hand, research tends to be developed in a temperature range such that ordering phenomena can be investigated. Under these conditions of lower temperature, kinetic factors may arise which make the approach to true equilibrium more difficult. Indeed, inside a solid particle, homogeneity of

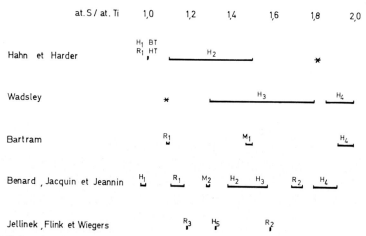

Fig. 2.7. Results obtained in the titanium–sulphur system by different authors. R, H and M indicate rhombohedral, hexagonal and monoclinic sulphide. A number indicates a particular unit cell. * indicates a region not studied.

composition is only ensured by diffusion. But diffusion coefficients are the smallest as preparation temperature is far from melting point. On the other hand, diffusion coefficients increase with deviation from stoichiometry, as shown in iron sulfide, $Fe_{1-x}S$, (Condit [1959]); in other words, a sulfide with ordered vacancies exhibits low autodiffusion. That might be the reason why some iron alloys are protected against H_2S corrosion, because they are covered by a continuous film of ordered Fe_9S_8 or Fe_7S_8 (Herzog [1959]; Meyer et al. [1958]).

If, for example, the titanium–sulfur system, studied independently in several laboratories in the past ten years, and already presented by Kjekshus and Pearson [1964], is considered, one is immediately impressed by the scattering of results (Fig. 2.7). All these sulfides have been prepared by allowing sulphur vapour, excluding all other gases, to attack titanium metal, which was either in a powdered or in a spongered state. Obviously, thermodynamic equilibrium is not easy to attain. Under these conditions,

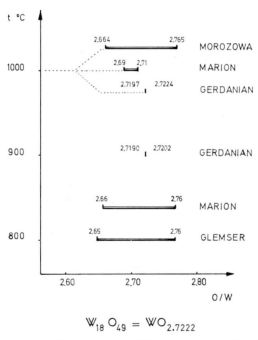

$$W_{18}O_{49} = WO_{2.7222}$$

Fig. 2.8. Nonstoichiometry of tungsten oxide $WO_{2.72}$ as determined by different authors.

the homogeneity of the products so prepared may be questioned. Illustrating this remark, let us mention the following experiment: if a mixture of sulphur and titanium in an atomic ratio close to 1.65 is heated in an evacuated sealed tube, sulphide is formed as powder and as single crystals; separate chemical

References p. 126

analysis of powder and crystals has shown that their composition is not the same.

Another interesting example is given by the oxygen–tungsten system around the $WO_{2.72}$ composition. Several authors, using different methods of preparation, have found a homogeneous phase whose width appears to differ from one study to the other, as shown in Fig. 2.8 (Morozowa and Guetzkina [1959]; Glemser and Sauer [1943]; Marion and Choain-Maurin [1962]). Again, it may be thought that equilibrium is not easy to attain. Work by Gerdanian and Marucco [1966] has, in the opposite way, revealed a very narrow phase. The equilibrium was obtained by putting the prepared oxide and another oxide in two separate boats; the second oxide was WO_3 or a mixture $WO_2 + WO_{2.72}$ depending on the desired limit of the nonstoichiometric field. Equilibrium was reached through the CO/CO_2 vapour phase and checked carefully by weighing the prepared $WO_{2.72}$ oxide. If this result is collated with the crystallographic work of Magneli [1949] on $W_{18}O_{49}$-ordered tungsten oxide, it may be concluded that the narrowest phase observed represents the best result, that is the actual state of equilibrium, insufficient carefulness yielding metastable states leading to an excessively wide homogeneous phase.

Quite often measurements are done on nonstoichiometric compounds at room temperature, when preparation involves a high temperature followed by cooling. The speed of cooling appears to be a non-negligible factor: this has been shown by Grønvold et al. [1958] for the vanadium–

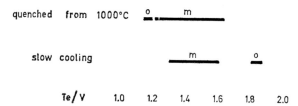

Fig. 2.9. Observed nonstoichiometry of vanadium tellurides when samples are quenched or slowly cooled. m and o stand for monoclinic and orthorhombic.

tellurium system. Fig. 2.9 shows some of their results. From them it is clear why it is necessary to quench samples, if solid state transformations are likely during cooling.

A similar remark may be made about iron monoxide. Its nonstoichiometry has been the subject of variable and even sometimes contradictory results. The actual nature of defects is due to iron vacancies (Jette and Foote [1933]) with simultaneous jumps of Fe^{3+}-cations in interstitial positions forming magnetite clusters (Roth [1960]). This oxide is not thermo-dynamically stable below 570°C, a temperature under which it decomposes into iron and magnetite. Manenc [1967] has shown that cooling must be very quick and very efficient to ensure ferrous oxide is kept free of any magnetite contamination. Thus the work of Roth could have been questioned on whether the described situation is or is not a high-temperature situation frozen by quenching, were his results not consistent with interpretations and results of Himmel et al. [1953] on iron self-diffusion in ferrous oxide.

On the other hand, Manenc et al. [1962], Manenc and Herai [1963], studying decomposition of this oxide, identified and then described three steps in this process. The first one is a pre-precipitation proposed because, at the first commencement of decomposition, satellites appear in the vicinity of characteristic Bragg reflexions of the FeO matrix. They are the result of a planar disorder in the lattice because platelets of a composition different from that of the matrix are formed. They might be an oxide richer in oxygen than the matrix; this oxide is metastable and its crystallization state may be described as a superstructure. This is not observed if experiments are not carried out with great care.

Nevertheless, in spite of care during cooling, it may happen that phase-change in solid state is so quick that it is necessary to make measurements at high temperature. An example is given by the determination of the width of the nonstoichiometric Ni_3S_2 phase, which is close to 0 below 550°C and very wide above this temperature, since it extends from $Ni_{2.56}S_2$ to $Ni_{3.68}S_2$ at 640°C. It has only been possible to measure this with the help of a high temperature X-ray camera (Huber and Liné [1963]).

A last experimental detail, which may have unwanted and unexpected consequences, is the behaviour of the reaction vessel towards the compound to be prepared: obviously it must be inert. Thus Hahn et al. [1957] thought they had prepared a zirconium sulfide, Zr_4S_3, after long treatment in a silica tube at high temperature of a mixture of sulphur and zirconium. Actually, as Hahn himself and Jellinek [1962] have pointed out later, silica went into reaction with zirconium and the product obtained was a thiosilicide ZrSiS.

The same phenomenon appears on a different scale when titanium sulfides whose composition is close to TiS are prepared. Thus Hahn and Harder [1956] thought they had prepared an ordered, high-temperature

References p. 126

TiS (titanium monosulfide), containing 9 times the basic unit as compared with two units in B8 TiS obtained at low temperature. Actually it has been shown by Bartram [1958] that this sulfide was a Ti_8S_9-ordered vacancy sulfide and by Jacquin and Jeannin [1963] that this nonstoichiometric phase was obtained when an approximate initial composition $S/Ti = 1.0$ was put into reaction. Some titanium has reacted with silica yielding a titanium silicide Ti_5Si_3, already found by Hahn and Harder, and the sulphide obtained was found to be sulphur-enriched. This phenomenon occurs every time when metal can react with silica, either when metal is in a low oxydation state, or when it is very reactive towards silica.

These remarks show how important it is to check and to describe the conditions of preparation. Otherwise, excessively large homogeneous metastable phases may be observed, high temperature structures may not be quenched, an intermediate phase may appear because of too slow cooling, or a refractory may react with the prepared compound.

2.3. Width of homogeneous phases

A difficult problem is the knowledge not only of factors which determine the width of nonstoichiometric phases, but also their range of influence, the way they act and their correlated influence. In the present state of the problem, some information has been obtained, but the widths of homogeneous phases cannot be precisely foreseen. They are extremely variable since cadmium oxide only contains a very small number of defects while the nonstoichiometric nickel telluride phase is made of all compounds whose composition ranges between $NiTe_{1.20}$ and $NiTe_{2.00}$ (Westrum et al. [1961]). Even more, three compounds of expected similar behaviour such as MnO, FeO and CoO, exhibit different properties since the lattice of the first one tolerates moderate deviations, $1.0 \leqslant O/Mn \leqslant 1.045$, the second large deviations, $1.05 \leqslant O/Fe \leqslant 1.13$, and the third one a narrow field of homogeneity.

However, some remarks may be made immediately. Very strongly ionic or very strongly covalent compounds, not containing any transition metal, able to exist in different oxydation states, admit of a very small number of defects. This is the case with sodium chloride or silicon carbide. If transition metals are considered, the situation is much more complicated. Some compounds have a very narrow homogeneity field, such as the $W_{18}O_{49}$ (tungsten oxide) already mentioned. Some others have a very broad field of homogeneity. This is the case with cerium oxide, including all compounds between CeO_2 and $CeO_{1.66}$ (Bevan and Kordis [1964]). This is true not only for oxides, but also for sulphides. For instance, PtS (platinum sulfide) (Grøn-vold et al. [1960]; Kjekshus [1966]) has a homogeneity range so reduced that it has not been detected; but TiS_2 (titanium disulphide) extends from 1.81 to 1.94 if the sulphides are made at 800°C (Jeannin and Bénard [1959]). Other examples can be found among nitrides, carbides, or even ternary compounds. VC_x is characterized by an x value changing continuously from 0.68 to 0.875 (Froidevaux and Rossier [1967]) while chromium carbides $Cr_{23}C_6$, Cr_7C_3 and Cr_3C_2 seem to be stoichiometric (Hansen [1958]).

The extent of a homogeneous nonstoichiometric phase depends on many variables including dimensional factors and the nature of the chemical bond between combined elements, which itself depends on external electronic structures or on electronegativities. To define their respective influences needs great care since these characters are usually strongly correlated. More-over, discontinuous changes are involved when, in a compound, one element is changed into another, either from the same column of the periodic table, or from a neighbouring atomic number.

The effect of atomic size factors can be demonstrated with the help of the following examples. Obviously they must involve point defects whose existence and number are strongly directed by such a size factor: they are substitutions or insertions. In order to investigate this effect, an isostructural series has first to be found.

The size factor is particularly well demonstrated by "bronzes". This kind of structure is known to accept interstitial ions such as those of alkaline metals, because empty space is provided by the lattice. Tungsten bronzes are very well known. Recently, vanadium bronzes have been thoroughly in-vestigated by Hagenmuller and co-workers. They have shown that the maximum amount of inserted alkaline metal is in close relation with its ionic radius. Thus the limit of solubility is 13% for lithium (Hardy et al. [1965]), 2% for sodium (Pouchard et al. [1967]), and only 1% for potassium (Pou-chard and Hagenmuller [1967]). The same is true of divalent cations, since the limit of solubility is 11% for magnesium, 3% for zinc, and 2% for cadmium (Hagenmuller et al. [1966]). When the attempt is made to insert more than the above-mentioned limits, a new phase occurs, called β-phase, whose structure has been determined by Wadsley [1955]. It is simply described as $(V_2O_5)_n$ chains with sodium, or more generally alkaline ions inserted between them in a nonstoichiometrical way. The extent of this β-phase is also closely related with the ionic radius of the inserted alkaline metal. Thus limit values are $0.22 \leqslant x \leqslant 0.62$ for lithium (Hardy et al. [1965]),

References p. 126

$0.22 \leqslant x \leqslant 0.40$ for sodium (Pouchard et al. [1967]), and $0.19 \leqslant x \leqslant 0.27$ for potassium (Pouchard and Hagenmuller [1967]) if the formula is written $M_x V_2 O_5$. The extent of these insertion phases decreases with increasing ionic radius of the alkaline metal inserted.

Another quite demonstrative example is obtained from mixed oxides of zirconium and rare earths $ZrO_2 - M_2 O_3$. Indeed rare earths provide a unique series because their electronic structure and electrochemical properties are quite similar and the only factor which shows appreciable change within this series is the ionic radius, that is a dimensional factor. These nonstoichiometric phases are of fluorite type. For the particular composition $2 ZrO_2 - M_2 O_3$, an order–disorder transition occurs (Perez y Jorba [1962]) provided that the ratio rare earth radius/zirconium radius is over 1.2. At transition temperature, both oxygen vacancies and Zr and M cations are ordered. This ordered phase, of pyrochlore structure, itself admits of deviations from stoichiometry. Two features of these phases are closely related to ionic radii. The first is the

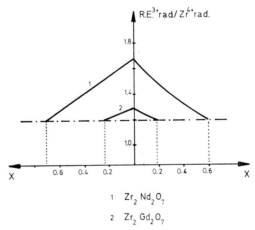

Fig. 2.10. Nonstoichiometry of $Zr_2 M_2 O_7$, where M is a rare earth, deduced from the variation of the ratio (rare earth ionic radius/zirconium ionic radius); this ratio must be greater than 1.2. (By courtesy of Collongues and of Silicates Industriels.)

transition temperature which will be discussed later (p. 123). The second is the extent of the nonstoichiometric $Zr_2 M_2 O_7$ phase. As a matter of fact nonstoichiometry is produced on both sides of this particular composition by substitution, yielding $(Zr_{2-x} M_x) M_2 O_7$ and $Zr_2 (M_{2-x} Zr_x) O_7$. In the ratio of ionic radii described above, following the case of substitution, the M or

Zr radius is replaced by an average radius, linear function of M and Zr radii and of x. Plotting the above ratio as a function of x allows as to determine the extent of the nonstoichiometric phase by crossing curves with a horizontal line corresponding to the particular ratio 1.2 (Fig. 2.10). A good agreement has been found between experimental values and the curves so constructed (Perez y Jorba et al. [1962]). For example, the observed domain of $Gd_2Zr_2O_7$ extends from 29.5 up to 37.9 mole per cent of Gd_2O_3 while calculated limits are 28.7 and 37.5; agreement is as good for $Nd_2Zr_2O_7$ whose observed domain extends from 19.1 up to 48.1 mole per cent of Gd_2O_3 while calculated limits are 18 and 48.

Another interesting family in which size factor can be important is that of transition metal chalcogenides. Many dichalcogenides crystallize into the CdI_2 type which is closely related to the NiAs type shown in Fig. 2.11. In both cases, metal surrounding is octahedral. This one of sulphur is strongly asymmetric in the C6 (CdI_2) type; in the B8 (NiAs) type it occurs as a trigonal

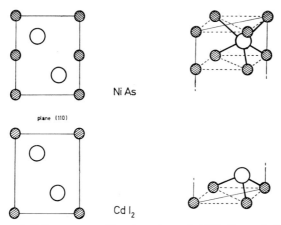

Fig. 2.11. NiAs and CdI₂-structures. The plane shown is (110) diagonal plane. The right-hand part represents the environment of the nonmetallic atom, white circle, while the metal atom is represented by a dashed circle.

prism. Therefore the C6–B8 continuous transition is possible through a gradual filling of the intermediate plane, empty in C6, fully occupied by metal in B8. Actually this seldom occurs: the zirconium–tellurium (Hahn and Ness [1959]) or nickel–tellurium systems (Westrum et al. [1958]) are good examples, although in the first case the existence of an ordered compound is suspected around $ZrTe_{1.3}$, and although in the second case transition is not quite complete. Much more frequently in this composition range which extends from MX to MX_2, multiple unit cells, which will be discussed in the second part, occur, because of vacancy ordering. However, when continuous transition is not observed, it may be thought that vacancy ordering is allowed by temperature conditions.

When the CdI_2 type is considered, the empty intermediate plane can receive additional atoms, either from the metal M of the binary MX_2 compound, or even from a different foreign element, yielding nonstoichiometry. For instance, the nickel insertion into nickel telluride, $NiTe_2$, lattice gives compounds of general formula $Ni_{1+x}Te_2$ and the extent of the observed homogeneous phase is quite large (Westrum et al. [1958]); the insertion of alkaline metals into the lattice of TiS_2 also being possible in the same way, giving a compound M_xTiS_2, as shown by Rüdorf [1965], and more thoroughly studied recently by Le Blanc et al. [1968]. In accordance with the nature of the inserted metal, its distribution in the intermediate plane is variable. In $Ni_{1+x}Te_2$ additional nickel atoms are inserted in the empty plane, while the other basal planes remain fully occupied. In K_xTiS_2, compounds similar to alkaline metal-graphite compounds are obtained; for $0.06 \leqslant x \leqslant 0.08$, one intermediate layer in four is occupied partially by additional potassium atoms; if the content in potassium is higher, that is if $0.14 \leqslant x \leqslant 0.16$, one in two layers contains additional potassium atoms. Thus electronegativity appears as a very effective factor, indeed.

Incidentally, it is worth pointing out that any compound of C6 type containing point defects does not necessarily tolerate insertion in the empty half *c*-height plane, although insertion indeed is the most frequently met situation. Thus is zirconium diselenide: it has been found that defects are substitution of selenium by zirconium together with zirconium vacancies (Gleizes and Jeannin [1969]).

To close these remarks about size factor, let us note that it is not because empty space is available that nonstoichiometry must occur. Thus, some dichalcogenides, although belonging to the C6-type do not accept a large number of additional atoms in the intermediate plane: it is platinum dichalcogenides whose nonstoichiometry has not been chemically detected (Westrum et al. [1961]; Grønvold et al. [1960]). The same is true for other lattice types, such as Nb_3S_4 (Ruysink et al. [1968]). The packing of atoms is such that channels are built by sulphur atoms which surround niobium atoms octahedrally. These channels are empty and they do not seem to accept additional atoms since this sulphide does not show noticeable deviations from stoichiometry, although channels offer quite sufficient space for interstitial niobium atoms. Again, this clearly shows that the size factor is not the only one to be considered when explaining the entrance of additional atoms, and that the nature of the chemical bond must largely influence deviation from stoichiometry. However, its influence is not always easy to display as size factor and chemical bond factor are obviously correlated.

In this respect, an interesting discussion results from comparing the plottings of the respective widths of nonstoichiometric phases crystallizing in strictly B8 and C6-types (Figs. 2.12 and 2.13). Firstly it may be noted that C6-type phases are on the average broader than those of NiAs-type. Secondly, from the inspection of these figures, in some cases, phase width is increased when sulphide is compared to selenide and to telluride. This might be attributed to a change in electronegativity difference, the broadening of the phase being bound up with an increase in metallic character. Nevertheless,

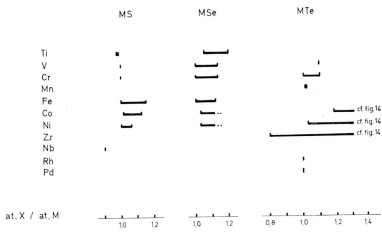

Fig. 2.12. Nonstoichiometry of compounds crystallizing in NiAs-type for various transitional-metal chalcogenides.

this remark does not appear general. It may be pointed out that comparable results have seldom been obtained by the same research group and it is explained in the second part that experimental data can sometimes be discussed because equilibrium is not easy to attain. Moreover, to build Figs.

References p. 126

2.12 and 2.13, results related to different preparation temperatures have been gathered, as a complete set of data is not yet available.

Considering B8 type compounds, a bonding type intermediate between ionic and metallic has been proposed several times, particularly by Raynor

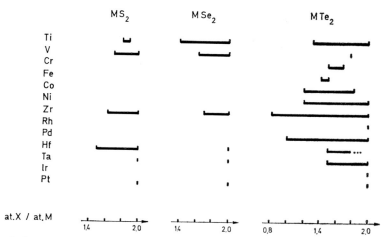

Fig. 2.13. Nonstoichiometry of compounds crystallizing in CdI$_2$-type for various transition-metal chalcogenides.

[1959] and by Mooser and Pearson [1959]. Ionicity is strongest when the c/a ratio is highest. Following a computation by Zemann [1958], $c/a = 1.77$ is the most favourable value for ionic bonding. Crystallization in NaCl type, is however, not likely because of the excessive difference in ionic radii of metal and chalcogen. On the other hand, low values of c/a introduce interactions between metallic atoms located one above the other along the symmetry axis, so that the metal coordination becomes eight. From this it may be inferred that some metallic bonding is then involved. Raynor justified this view about ionicity considering together variations of c/a ratio and of formation enthalpy, which may be understood as an approximate measure of the degree of electrochemical interaction. Compounds of low c/a, the most metallic ones, show the largest deviations from stoichiometry. Compounds of high c/a, the most ionic ones, only admit small deviations (Table 2.1) and exhibit a pronounced tendency to the vacancy ordering when the X/M ratio increased. As a matter of fact, Bertaut [1953] has established that this phenomenon of ordering may be explained using a purely ionic model for the chemical bond mechanism. Lattice energies of ordered and disordered iron sulphide, Fe_7S_8, have been computed, from this it is found that the ordered state is more stable. The ordering energy found is much too high to be accepted; obviously a purely ionic model is not entirely valid; however,

Table 2.1

MX	c/a	ΔH^0 kcal/mole	Domain at. %
VS	1.748	−22.5	?
FeS	1.63	−11.4	3.3
MnTe	1.618	−11.25	0.5
NiS	1.555	−14.0	1.46
CoS	1.538	−11.0	3
NiSe	1.461	− 5.0	?
CoSe	1.463	− 5.0	?
CoTe$_{1.2}$	1.381	− 4.5	9.7
NiTe	1.350	− 4.5	16.1

Bertaut has concluded that, even if chemical bond is partially ionic, this ordering energy is high enough to allow an ordering of vacancies at low temperature.

Thus chemical bonding is an important factor which actually covers some other fundamental reasons, among which may be quoted external electronic structures, or electronegativities of the combined elements. Since they are interrelated, their own influence is difficult to describe, the more so as variation is discontinuous when compounds are made with different chemical species. A continuous change can be observed when there is a solid solution. This is the case with the system $(Fe, Mn)_{1-x}O$. The first remark to be made is related to the respective breadths of the nonstoichiometric FeO and MnO phases; the first is broad, and the second is narrow. The similarity in crystal structure for both compounds clearly indicates the considerable influence of electronic configurations. A continuous change in the number of defects in chemical equilibrium is observed when the ratio Mn/Fe is varied (Voeltzel and Manenc [1966]) (Fig. 2.14). The same is true for chalcogenides, such as, for instance $Ti(S, Se)_2$; but in this case the width of the nonstoichiometric field does not change regularly with the ratio S/Se (Jacquin and Huber [1966]) (Fig. 2.15).

References p. 126

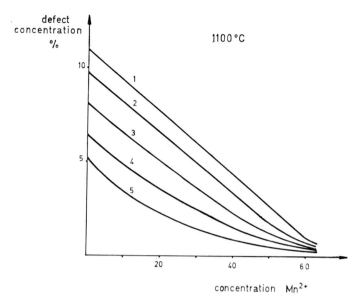

Fig. 2.14. Concentration of cationic vacancies in (Fe, Mn) O as a function of manganese cation concentration. Curves are related to different ratios p_{CO}/p_{CO_2}: 1 — 0.39, 2 — 0.59, 3 — 1.13, 4 — 1.86. Curve 5 is related to an iron-saturated oxide. (By courtesy of Voeltzel and Manenc and of Gauthier-Villars.)

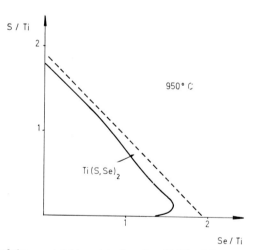

Fig. 2.15. Width of the nonstoichiometric domain of MX_2–CdI_2 type for solid solutions TiS_2–$TiSe_2$. (By courtesy of Huber and Jacquin and of Gauthier-Villars.)

The study of a nonstoichiometric compound cannot be accepted as complete if the nature of defects is not determined. For narrow non-stoichiometric phases, particularly those which exhibit semiconducting, type of semiconductivity can be used. For grossly nonstoichiometric phases, the density method is still frequently applied. The principle is simple. Measured value is compared to computed values, considering three elementary assumptions: vacancies, inserted atoms, antistructural defects. As an example, curves related to TiS_2 phase clearly show that additional titanium atoms are responsible for deviation from stoichiometry (Fig. 2.16) (Jeannin and Bénard [1959]).

Such a description is here considered as oversimplified. Indeed, the experimental curve lies systematically below the theoretical curve correspond-

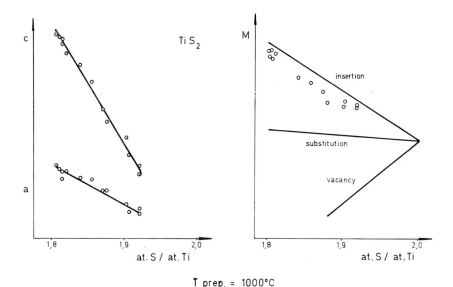

T prep. = 1000°C

Fig. 2.16. Density measurements combined with lattice constant measurements allow of computation of mass contained in the unit cell; from comparison with theoretical computations the nature of defects can be deduced.

ing to the chosen type of defect. Moreover, such a deviation cannot be understood as an experimental error. The curve related to titanium disulfide already shows a small deviation. This deviation is larger for nonstoichiometric bismuth telluride (Miller and Che Yu Li [1965]) at large deviations

References p. 126

from stoichiometry. It is even more important for titanium monoxide, even for chemically stoichiometric composition since deviation is as high as 14%. This last compound has been studied independently by Andersson et al. [1957] and by Straumanis and Li [1960]. Although experimental values of limits of the TiO homogeneous phase differ somewhat since Andersson et al. have found 0.64 and 1.26 while Straumanis and Li propose 0.86 and 1.20, the differences between calculated and observed densities are the same and cannot be questionned. To explain this fact, it has been suggested that this systematic difference may be due to the simultaneous presence of metallic and nonmetallic vacancies (Arija and Popov [1962]).

Starting from this idea of the coexistence of two types of defects, Van Gool [1966] has made some computations yielding the following results. In titanium disulfide there are some sulphur vacancies in a small number at the same time as interstitial titanium atoms: thus in a sulphide of composition $S/Ti = 1.900$, prepared at $1000°C$, there should be 0.057 titanium atoms inserted and 0.018 sulphur sites vacant. In titanium monoxide, titanium and oxygen vacancies exist simultaneously. In fact, in the case of an atomic ratio corresponding to stoichiometry, 15% of the lattice positions are unoccupied (Table 2.2). This remark must be considered alongside the fact that,

Table 2.2

	Ti vac.	O vac.	total
$Ti\ O_{1-0.284}$	6.1%	32.8%	19.5%
$Ti\ O_{1-0.140}$	10.5%	23.1%	16.8%
$Ti\ O_{1-0.108}$	10.8%	21.3%	16.0%
$Ti\ O_{1-0.030}$	14.3%	16.8%	15.6%
$Ti\ O_{1-0.005}$	14.6%	15.0%	14.8%
$Ti\ O_{1+0.060}$	17.2%	12.2%	14.7%
$Ti\ O_{1+0.116}$	18.5%	9.1%	13.8%
$Ti\ O_{1+0.150}$	20.1%	8.2%	14.2%
$Ti\ O_{1+0.195}$	21.2%	5.8%	13.5%
$Ti\ O_{1+0.200}$	22.1%	6.5%	14.3%
$Ti\ O_{1+0.250}$	23.4%	4.2%	13.8%

according to Straumanis and Ariya, TiO is made up of titanium atoms in two states of oxidation, both different from value two. Such a simultaneous presence leads to short-range ordering, which will be discussed later. Nevertheless, it is worth pointing out that this assumption, of two metallic oxidation states coexisting, is not supported by the observed metallic conductivity of titanium monoxide.

This large difference between observed and calculated densities for the chemical stoichiometric composition does not occur in every case. For instance, the density curve of bismuth telluride as a function of composition shows the more serious discrepancy as deviation from stoichiometry increases (Miller and Che Yu Li [1965]). With stoichiometry, theoretical and computed values are equal, within the limits of experimental error.

The two point defect types can be of the same type, as in titanium monoxide, but also of different types, as in mixed oxide made of zirconium dioxide and calcium oxide, studied by Carter and Roth [1967]. Lattice is of fluorite type. Density of a nonstoichiometric oxide can be computed either assuming interstitial zirconium and calcium atoms, both species being statistically distributed, or assuming oxygen vacancies. For an oxide containing, for example 14.2% mole per cent calcium oxide, the observed density is 2% higher than the value computed for oxygen vacancies. Since this difference is quite considerably higher than experimental error on density, it is thought to be due to cations in interstitial positions, as allowed by fluorite structure. Thus, for a sample quenched from 1900°C containing 14.2 mole per cent calcium oxide, 19.7 unit cells contain interstitial cations while 80.3 unit cells have vacant anionic sites.

3. Ordered phases

3.1.

A small deviation from stoichiometry, sometimes so small that it can escape chemical experimental detection, is due to a low number of point defects. Because of this low number, they can be assumed without interaction; in other words, the distribution of defects is purely statistical. The situation is identical to that of a dilute solution. Physico-chemical properties are then interpreted using theoretical conclusions of Wagner and Schottky [1931]. However, when the number of defects introduced in a lattice grows, interaction between defects becomes more important. Then interaction energy can no longer be neglected. Such interactions give rise to particular arrangements of point defects on a more or less important scale.

The smallest interaction obviously is that between two defects; an

References p. 126

example has been met earlier with respect to digenite. But interaction can extend over a few unit cells of the basic disordered system, yielding "submicrodomains". Therein, atoms stay very close to the positions that they occupy in the disordered state. But these submicrodomains do not necessarily have the averaged composition given by chemical analysis. The system is then microheterogeneous, being made up of submicrodomains of two kinds, different in composition and structure. These submicrodomains are coherent, which means that misorientation between them is very small (Wadsley [1955]; Ariya and Popov [1962]). As a final state, the ordering of defects arises on the scale of the whole crystal.

Examples of these different possibilities will be discussed successively.

3.2. Long-range ordering

When the study of nonstoichiometry was in its initial stages, many compounds were said to have a broad homogeneity range. At present many such phases have been resolved in a sequence of ordered phases, each having its own much more limited nonstoichiometry. For instance, Ehrlich [1939] proposed for Ti_3O_5 a large homogeneous domain, including all compositions defined by atom ratio $1.65 \leqslant O/Ti \leqslant 1.90$. Actually, following results of Andersson et al. [1957], it is accepted that this range comprises a series Ti_nO_{2n-1}, made of the successive oxides Ti_3O_5, Ti_4O_7, Ti_5O_9, Ti_6O_{11}, Ti_7O_{13}, Ti_8O_{15}, Ti_9O_{17}, $Ti_{10}O_{19}$. Minimum free enthalpies of these oxides must be very close indeed and special care must be taken in isolating them during preparation. This phenomenon appears general since it has been observed not only for transition metals compounds but also for semiconducting compounds exhibiting large deviation from stoichiometry. Brebrick [1969] has described a continuous sequence of phases for the Bi–Te system in the 50–57 at.% Te region if compounds are synthetized at 450°C. It was described before as a broad homogeneous range. The hexagonal unit cells, characterizing each phase, have the same basal plane while the hexagonal axis is a function of two integers, depending on composition and on annealing temperature.

Ordered phases are not always easy to observe. As a matter of fact, it is often necessary to carry out the preparatory heat treatment at a not too high temperature to avoid complete disorder, and thus synthesis may take a long time. Moreover, ordering goes with the occurrence of multiple unit cells, which may be very large in some cases, and which may sometimes be of a lower symmetry. This makes the study of such ordered phases difficult. Of course an ideal situation occurs when an ordered single crystal is prepared. But then accurate chemical analysis of such a crystal sets a hard problem. Thus, in some cases the formula is straightforwardly deduced from X-ray work and then checked by direct synthesis with the assumed composition which must yield a homogeneous phase with the expected structure.

Elimination of randomly distributed point defects may be made in two ways, depending on the type of compounds. Two occur particularly frequently. Either point defects, vacancies or interstitials, get ordered, that is periodically distributed, or a more complex process, called shearing, is involved. After vacancies have been regularly arranged, oxygen atoms move to occupy these empty positions, carrying a slice of the lattice in such a way that metal atoms contained in this slice have jumped into interstitial positions again finding a six-coordination environment.

But some other particular orderings are likely, such as that described by Carter and Roth [1967] of calcium-stabilized zirconia. In this case already mentioned, oxygen presents a distorted statistical tetrahedral environment of a normal site, which becomes ordered after long annealing at $1000°C$. Strictly speaking it is not a defect ordering, but a consequence of a particular defect situation.

It is interesting to point out here that growing conditions of ordered domains may hinder long-range ordering when this ordering is accompanied by only very slight deformations. To this repect, the work of Belbeoch et al. [1967] of uranium oxide, U_4O_9, is quite interesting. This oxide containing interstitial clusters is cubic at high temperature; it gets ordered below $65°C$ then crystallizing into a rhombohedral system. However, this reversible transformation may not be detected, the reason being that, during cooling, the nucleation of small rhombohedral domains occurs along (111) cubic lattice rows. But in such a cubic crystallite there are four (111) equivalent directions; a uniform distribution between the four possible orientations is possible. Elastic shear constraints maintain the coherence of these rhombohedral domains such that the unit cell remains isometric. Ordering cannot be observed macroscopically by X-ray methods. However, if the size of initial crystallites is very small, only one rhombohedral domain grows. It has been experimentally shown by appropriate heat treatments that the critical size of the crystallite is 0.1. This experiment is important because it stresses the importance of the thermal history of the sample which controls grain size, to the scale of ordering.

References p. 126

Ordered vacancies

Ordered-vacancy compounds may be divided into two groups, those with anionic vacancies, and those with cationic vacancies.

Ordered anionic vacancies

Examples of compounds with anionic ordered vacancies may be often found among rare-earth oxides or mixed oxides containing rare earth which derive structurally from fluorite types.

For instance, zirconia crystallizes into a fluorite system when it is associated with various other oxides, such as rare-earth sesquioxide M_2O_3. Actually one gets a nonstoichiometric compound of which the right formula is $Zr_{2-2x}M_{2x}O_{4-x}$, since the anionic oxygen sub-lattice contains oxygen vacancies. For $x=0.5$, that is for $Zr_2M_2O_7$, an order–disorder transition is observed and a pyrochlore structure is found (Perez y Jorba et al. [1959]). It must be pointed out that in this particular case, not only vacancies but also cations are ordered.

More complicated examples are given by rare-earth oxides or by some mixed oxides of a ratio different the above one or even by mixed oxides,

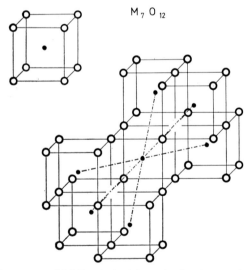

$M_7 O_{12}$

Fig. 2.17. Crystal structure of M_7O_{12} derived from the fluorite type by means of ordered oxygen vacancies. Cations are six-coordinated or seven-coordinated by oxygen anions. (By courtesy of Sawyer, Hyde and Eyring and of Société Chimique de France.)

(a)

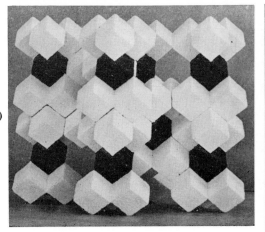

b)

(c)

Fig. 2.18. Mixed oxides of zirconium and scandium, of fluorite type, can be described as arrangements of blocks shown in (a). If only (a) left blocks are assembled, one gets $Zr_3Sc_4O_{12}$ (M_7O_{12}) (b). If left and right in (a) blocks are together in one to one ratio, $Zr_5Sc_2O_{13}$ (c) is obtained. (By courtesy of Thornber, Bevan, Graham and of Acta Crystallographica.)

involving oxides such as MO_3 where M is molybdenum, uranium or tungsten (Bartram [1966]). For instance, Fig. 2.17 shows the ordering of oxygen vacancies in an oxide of formula M_7O_{12}. In the compound studied by Baenziger et al. [1961], (M = Tb) the vacancy arrangement is such that one metallic atom is surrounded by six oxygen atoms and six metallic atoms by seven oxygen atoms. Six-coordinated cations are arranged along an infinite string which is the ternary axis of the fluorite basic cell, and this string is surrounded by seven coordinated cations. Anion vacancies are regularly distributed along this axis when ordered; no vacancies correspond to fluorite structure. M_2O_3 and M_7O_{12} where M is a rare-earth element are understood as different ways of interweaving these strings (Hyde and Eyring [1965]; Hyde, Bevan and Eyring [1965]). From this base, they have been proposed as basic structural units for the series Me_nO_{2n-2}, to which Pr_9O_{16} and $Pr_{12}O_{22}$ belong, and for which models have been suggested (Sawyer et al. [1965]).

However, the discussion by Thornber, Bevan and Graham [1968] of results obtained on structures of $Zr_5Sc_2O_{13}$ and $Zr_3Sc_4O_{12}$ does not strictly support this view. Basic units are rather better described as two groups of seven cubes; the first one consists of a six-coordinated cation surrounded by seven-coordinated cations, the other of seven eight-coordinated cations. It should be noted that strings mentioned above are easily found in units of the first type. In both cases, $Zr_5Sc_2O_{13}$ and $Zr_3Sc_4O_{12}$, the idealized cubes shown in Fig. 2.18 share edges. Then M_nO_{2n-2} phases may be understood as arrangements by edge sharing of one unit of the first type with x units of the second type. For instance, $x=0$ corresponds to $Zr_3Sc_4O_{12}$ and $x=1$ to $Zr_5Sc_{12}O_{13}$. Moreover, nonstoichiometry may be explained on this basis (see p. 122), what makes this model more interesting.

Ordered cationic vacancies

Examples of compounds having cationic ordered vacancies appear quite frequently among transition metal chalcogenides. Generally met formulae are shown in the following table, where M is the metal and X the chalcogenide. These notations would almost bring us back to the simple idea of definite compound if the ordered compound itself did not admit of deviations from the above stoichiometric formula. Indeed they have not always been determined or even displayed. Probably more sensitive experimental methods would detect them. As a matter of fact, it is difficult to consider

Table 2.3

M_8X_9	$MX_{1.125}$	Ti_8S_9	Bartram [1958]
M_7X_8	$MX_{1.143}$	Fe_7S_8	Bertraut [1953]
M_5X_6	$MX_{1.200}$	Cr_5S_6	Jellinek [1957]
M_4X_5	$MX_{1.250}$	Ti_4S_5	Wiegers [1967]
M_3X_4	$MX_{1.333}$	Nb_3Te_4	Selte et al. [1964]
M_2X_3	$MX_{1.500}$	Zr_2Se_3	Mc Taggart et al. [1958]
M_5X_8	$MX_{1.600}$	Ti_5Se_8	Chevreton [1964]

that chromium sulphide, Cr_3S_4, (Jellinek [1957]) shows a homogeneous phase, going from $CrS_{1.266}$ to $CrS_{1.316}$ and that chromium sulphide, Cr_2S_3, has no stoichiometric deviation, when structures are closely related, and considering the thermodynamical remarks made earlier.

Structure of Fe_7S_8, described by Bertaut [1953], was the first example of vacancy ordered structure, which is closely related to the B8 type. Vacancies are regularly distributed in one of two metallic planes, the other being filled up. As a consequence of the ordering of vacancies the unit cell becomes a multiple of B8 unit cells, and a lowering of symmetry is involved.

Since the publication of that work, many distributions of cationic vacancies have been found. A particularly interesting family is the one based on B8 structure. Its basic unit may be described as an infinite sheet of chalcogenide octahedra sharing edges and containing a metal atom. Putting one over another one, sheets are separated by an empty plane in CdI_2 or MX_2, or by a plane fully occupied by metal atom in NiAs or MX. For intermediate compositions, these units are piled up in many different ways, which may yield many multiple unit cells as shown in the titanium sulphur system, where the *a* axis of the NiAs or CdI_2 unit cell stays the same, but the *c* axis, height of the basic unit, multiplied by an integer. The number of times an elementary sheet is taken is (Wiegers [1967]):

Table 2.4

2	TiS_2	BC	$P\bar{3}m$
2	TiS	BC	$P6_3/mmc$
4	Ti_2S_3	ABAC	$P6_3mc$
9	Ti_8S_9	CBABACACB	$R\bar{3}m$
10	Ti_4S_5	BCACBCBABC	$P6_3/mmc$
12	Ti_5S_8	BCABABCACABC	$R\bar{3}m$
21	Ti_3S_4	CBABCBABACABACACBCACB	$R\bar{3}m$

References p. 126

A, B and C stand for sulphur atoms located in the (110) diagonal plane; x and y coordinates are $(0,0)$ for A, $(\frac{2}{3}, \frac{1}{3})$ for B and $(\frac{1}{3}, \frac{2}{3})$ for C. Simultaneously with ordering, a lowering of symmetry may occur but this is not systematic, as shown by the last column of the preceding table which quotes space groups. It may be of different types. A change of sulphur packing with respect to NiAs may result in a modification from hexagonal to rhombohedral as for Ti_5S_8. But it can also be more important, particularly when the a axis is itself multiplied by an integer. Thus Fe_7S_8 is monoclinic (Bertaut [1953]) and Fe_7Se_8 is triclinic in its low-temperature form (Okasaki [1959]). However similarity with B8 remains considerable because the general frame of the structure is always due to the nonmetallic element not responsible for defects, and because chemical bonding does not change too much within this range of composition.

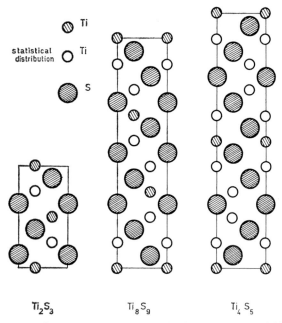

Fig. 2.19. Diagonal (110) plane, c axis being vertical, of three titanium sulphides containing ordered metallic vacancies. Ti_2S_3 contains one partially statistically occupied plane per fully occupied plane, Ti_8S_9 two partially per one fully occupied, Ti_4S_5 four partially per one fully occupied. For Ti_4S_5 the four statistically occupied planes do not present the same degree of occupation. (By courtesy of Jellinek and Wiegers.)

Casting aside lowering of symmetry as of secondary interest, the description of ordered cells of B8 type can be made by considering planes of compact hexagonal array of B8 type. Two features will be distinguished:
– the first is related to vacancy distribution in such metallic hexagonally arrayed planes,
– the second is related to the stacking of such planes, or rather, basic units described earlier.

As to the first point, vacancies may be
1-distributed randomly on the whole of metallic sites. This is the case of nickel sulphide (Laffitte [1959]);
2-localized in some planes, but distributed randomly in these particular planes: titanium sulphides are good examples: one plane containing vacancies per fully occupied plane is characteristic of Ti_2S_3 (Wadsley [1957];

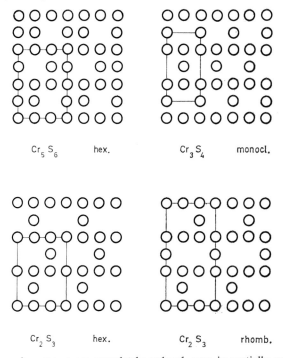

Fig. 2.20. Chromium atoms are completely ordered, even in partially occupied planes. Diagonal (110) plane is shown, c axis being vertical. (By courtesy of Jellinek and of Acta Crystallographica.)

Wiegers [1967]), two vacant for one occupied in Ti_8S_9 (Bartram [1958]) four

References p. 126

vacant for one occupied in Ti_4S_5 (Wiegers [1967]) (Fig. 2.19);
3-ordered completely. The best example is given by chromium sulphides, studied by Jellinek [1957]. Vacancy and filled up planes occur in turn (Fig. 2.20). In vacancy plane, one site in three is vacant for Cr_5S_6, one in two in Cr_3S_4, and two in three in Cr_2S_3. As a limit, it may happen that vacancies gather in the same plane till it is empty. C6-structure and Fe_3S_4 are examples (Erd et al. [1957]).

Next point is the stacking of hexagonally arrayed planes.
1. The first group of compounds keeps the sequence of nickel arsenide, that is:

... M   X type B   M   X type C   M   ...

if we consider a diagonal plane (110) of the basic cell. Fe_7S_8 or chromium sulphides are examples (Fig. 2.20)
2. The second group sees the sequence of nonmetallic atom more or less changed. Nevertheless, the environment of a metallic site, vacant or not, stays octahedral. Titanium sulphides are examples which show more or less complicated sequences, described earlier (Fig. 2.19).
3. The third group of compounds in which two nonmetallic planes of the same type follow one another in such a way that nonmetallic atoms are on the same vertical line, normal to the hexagonally arrayed plane. Thus, the environment of the metal becomes a trigonal prism. The simplest example is molybdenum sulphide where the sequence of nonmetallic planes is as follows:

... B B C C ...

This is C7-type, BB is called a sheet. Stacking faults of sheets may occur. Occurring regularly, unit cells multiple of C7 one are built. Examples are given by polymorphic tantalum diselenides (Jellinek [1962]; Brown and Beerntsen [1965]; Huisman and Jellinek [1969]), or by niobium sulfide (Jellinek [1962]), (Fig. 2.22).
4. The fourth group contains compounds for which the stacking is more complex, being simultaneously of B6 and C7 type; both kinds of metal surroundings, octaedral and trigonal prismatic, are found together. This may be the result either of nonstoichiometry in a chalcogenide belonging to the preceeding group, or of the slipping of C7 slabs. The nonstoichiometric effect is shown by niobium selenides (Jellinek [1962]), (Fig. 2.22). The slipping effect can also be described as regularly repeated stacking faults;

this is illustrated by some polymorphic forms of tantalum diselenide (Brown and Beerntsen [1965]; Bjerkelund and Kjekshus [1967]). Random stacking faults are also found in some cases, such as tantalum disulfide (Jellinek [1962]), or such as nonstoichiometric molybdenum disulfide (Mering and Levialdi [1941]).

This description seems simple but is limited to some chalcogenides of transition metals. As a matter of fact, it does not include all intermediate compounds of this class as, for instance, chalcogenides of niobium, such as Nb_3S_4, Nb_3Se_4, Nb_3Te_4, Nb_2Se_3, Nb_2Te_3, which cannot be classified following criteria used above. These structures are roughly described as made of indefinite, zigzag chains of niobium atoms which are octahedrally surrounded.

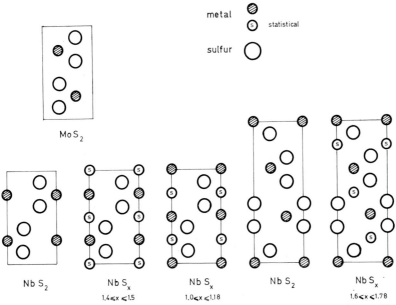

Fig. 2.21. Diagonal (110) plane, c axis being vertical, of different niobium sulfides. Niobium disulfide presents two types of crystallization. Nonstoichiometry is due to insertion of niobium atoms. Widths of nonstoichiometric fields are written below. The surrounding of niobium is either octahedral or trigonal prismatic. (By courtesy of Jellinek and of Elsevier.)

A distortion of these octahedras brings some niobium atoms so close to one another that interatomic metal–metal distances becomes quite similar to those observed in the metal, suggesting a metallic bond (Selte and Kjekshus

References p. 126

[1964]; Ruysink et al. [1968]). This is a feature already pointed out with respect to nickel-arsenide structure (p. 98). On the other hand, nonstoichiometry of some chalcogenides, and particularly sub-chalcogenides, cannot be understood on the basis of cationic vacancies. At least, there are compounds with cationic vacant sites which are not of nickel-arsenide type, such as transition metal oxides, for instance iron oxides.

Sheared structures

Till now, we have only discussed point defects ordering. But it also occurs that vacancies eliminate themselves and disappear through dislocations. An interesting discussion of this process has been published by Anderson and Hyde [1967]. This process would be a major one in generating what is commonly called shear structures, particularly found among the

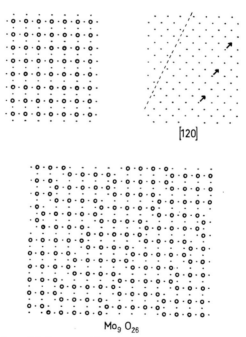

Fig. 2.22. Shearing in molybdenum oxide. The basal plane of rhenium trioxide structure is represented at top right. Top left shows only metal distribution with ordered vacancies arranged along [130]. Translation in the direction of arrow yields elimination of metal vacancies and then sheared structure appears. (By courtesy of Anderson and Hyde and of Pergamon Press.)

series of titanium oxides Ti_nO_{2n-1}, molybdenum oxides Mo_nO_{3n-1}, and tungsten oxides W_nO_{3n-2}, the study of which started under the impulse of Magneli. Following Anderson and Hyde, anionic vacancies revealed by experiment and due to the departure of oxygen from the lattice of the stoichiometric compound, gather in a well defined crystallographic plane. Then a lattice shift, eliminating vacancies when in the right number, again gives to the metal the 6-coordination. This suggests that covalency plays a by no means negligible part in the chemical bonding mechanism since metal atoms, as in coordination compounds, tend towards the six-coordinated octahedral environment. The region in which vacancies get together is limited by a dislocation ring. Its Buerger vector corresponds to an edge dislocation if it is normal to the ring segment, a screw dislocation if it is parallel, and a partial dislocation if it is not a lattice vector. Figs. 2.22 and 2.23 show the case of molybdenum oxide and of tungsten oxide. The formula of an oxide

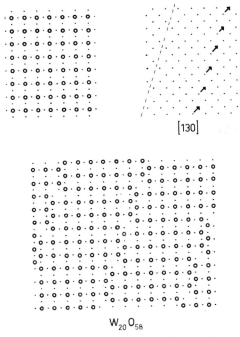

[130]

$W_{20}O_{58}$

Fig. 2.23. A different direction of alignment of ordered vacancies and a different distance between parallel directions of shearing yields different formulae for the sheared oxide. As an example $W_{20}O_{58}$ is shown. (By courtesy of Anderson and Hyde and of Pergamon Press.)

References p. 126

obtained in that way is ruled by the indices of the shear plane and by the distance between parallel shear planes.

Three elements help to define the shearing mechanism: the shear plane, the Buerger vector and the slip plane. For instance, for $W_{20}O_{58}$, an observed shear plane is $(3\bar{1}0)$ and the shear vector is $\frac{1}{2}a$ (110). For rutile derivatives, the situation is similar although a little more difficult to see, since the drawing implies three-dimensional space as shear planes may be $(1\bar{2}1)$ or $(1\bar{3}2)$ and the vector is $\frac{1}{2}(0ac)\,0\bar{1}1$. This interpretation is particularly fruitful because the nature of defects in TiO_2 was questioned. Thus, a small deviation from stoichiometry was interpreted by oxygen vacancies as suggested by experiments (Förland [1964]; Straumanis and Ejima [1961]). However, when TiO_2 is heated to a high temperature, the nature of defects appears in the form of interstitials of titanium, as shown by electron-spin resonance and by internal friction experiments (Chester [1961]; Carnahan and Brittain [1963]). This result is in agreement with shearing interpretation. Experimental evidence obtained with the electron microscope may support this mechanism (Ashbee et al. [1963]; Van Landuyt et al. [1964]). Multi-ordered sheared structures have been crystallographically studied; many of those have much more complicated formulae, they are binary or quaternary compounds. Obviously, they cannot be made experimentally starting from the ideal structure involving a coupling of reduction and shearing as for TiO_2 (Eikum and Smallman [1965]); but theoretically they can be described as a family through shearing. Thus, for instance, the series of alkali titanates whose end member has been found $Rb_xMn_xTi_{2-x}O_4$ (Fig. 2.24a) (Reid et al. [1968]). By shears, alkali titanates $Na_2Ti_6O_{13}$ (Fig. 2.24b) (Andersson and Wadsley [1962]) and $Na_2Ti_7O_{15}$ (Fig. 2.24c) (Wadsley and Mumme [1968]) may be obtained. Metal oxygen TiO_6 edge-sharing octahedra form puckered sheets between which alkaline cations are located. $Na_2Ti_6O_{13}$ is made of groups of three octahedra, and $Na_2Ti_7O_{15}$ of more complex groups of three and four octahedra; again alkaline cations are located between octahedra sheets, but in reduced number.

An equivalent description by octahedron blocks is met in some other systems, such as the WO_3-Nb_2O_5 system. Different chemical formulae are related to different sizes of octahedron blocks, while the array built by blocks is basically identical. $WNb_{12}O_{33}$ is made of blocks of three by four octahedra, four by four for $W_3Nb_{14}O_{44}$, four by five for $W_5Nb_{16}O_{55}$, five by five for $W_8Nb_{18}O_{69}$. $W_4Nb_{16}O_{77}$ is a even more complex case, since it is made of three by four blocks observed in $WNb_{12}O_{33}$, and of four by four blocks

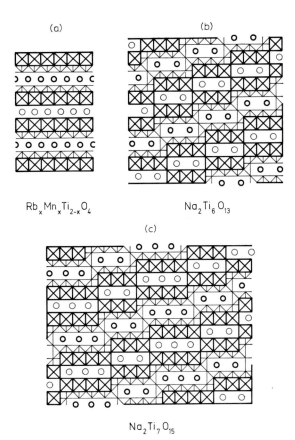

(a)

(b)

$Rb_x Mn_x Ti_{2-x} O_4$ $Na_2 Ti_6 O_{13}$

(c)

$Na_2 Ti_7 O_{15}$

Fig. 2.24. Shearing in ternary oxides containing alkaline metals. Such a family is shown, made of puckered sheets of octahedral TiO_6 sharing edges. They are regularly arrayed in (a) so that there is a continuous layer of alkaline metals, sheared in blocks of three octahedra in (b), and sheared in complex blocks of three and four octahedra in (c). (By courtesy of CSIRO and of Acta Crystallographica.)

observed in $W_3 Nb_{14} O_{44}$ (Andersson and al. [1966], Roth and Wadsley [1965]).

A review of structures containing shearing has been presentented by Wadsley [1964] and more recently by Andersson [1968]. Many types of such structures containing sheared slices, or sheared blocks, or sheared strings are known at present, which make the results of this mechanism general and which suggest another model for generating nonstoichiometry (p. 122).

References p. 126

3.3. Short-range ordering

Long-range ordering, although not always very easy to display, has nevertheless been extensively and precisely studied by X-ray methods, both in oxides and sulphides or other chalcogenides, or more complex compounds. The problem is quite different when short-range ordering is concerned. X-ray techniques are then not always sensitive enough to detect it. However some cases of short-range ordering, bound to defects interaction, have been recognized and clearly described. The interesting point is that one may start from this idea of short-range ordering, experimentally checked in some particular cases, and then in a convergent manner and at the same time, extrapolate these results and those of long-range ordering to put forward some new ideas on nonstoichiometry.

Short range ordering has been found in broad nonstoichiometric domains. Such a short-range order is already known in alloys, particularly in those exhibiting long-range ordering. Thus, it may be inferred that it should arise in grossly nonstoichiometric compounds, in which chemical interaction is stronger and more directed than in alloys. Several examples of different types will be described, leaving aside the case of digenite already discussed (p. 83).

Vacancy association can be demonstrated when the number of vacancies surrounding a vanadium atom in a nonstoichiometric vanadium carbide is determined. Its stability range extends between $VC_{0.66}$ and $VC_{0.875}$; this carbon-rich limit corresponds to the ordered V_8C_7 carbide. The type of crystallization of the nonstoichiometric phase is sodium chloride. Nuclear magnetic resonance technique can be applied to recognize the environment of vanadium because one vanadium isotope, $^{51}_{23}V$, has a nuclear spin equal to $\frac{7}{2}$. As a matter of fact, for a spin value greater than $\frac{1}{2}$, an electric field gradient split a Zeeman line because of its coupling with the quadrupolar moment of the nucleus. Yet ideal structure implies that every atom is octahedrally surrounded by six carbon atoms, some of those being missing because of nonstoichiometry. If one or more of these six carbon atoms is missing, the octahedral symmetry of surrounded vanadium atoms is destroyed, so that an electric field is induced. As the effect of the field decreases very quickly with distance, this method allows of the description of only the nearest neighbours of the vanadium atom. It is possible to associate each line of the resonance spectra with one or more types of "vanadium sites", this meaning a vanadium surrounded by a definite number of carbon vacancies. Figure 2.25 illustrates, for different carbon contents, relative probabilities of finding one vanadium atom surrounded by 0 vacancy or 6 carbon atoms, by 1 vacancy or 5 carbon atoms, by 2 vacancies or 4 carbon atoms and so on. Black marks represent experimental values. White marks are related to computed values, assuming a statistical vacancy distribution. This means that the formation energies of different "vanadium sites" are equal, or that no interaction energy occurs between vacancies. From this comparison, it is found that the actual vacancy distribution is never purely

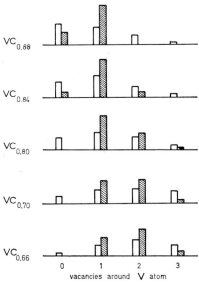

Fig. 2.25. Distribution of carbon vacancies around a vanadium atom for various vanadium carbides. White rectangles are related to computed values assuming purely statistical distribution of vacancies. Dashed rectangles are related to experimental values. Differences show that vacancies interact. (By courtesy of Froidevaux and Rossier and of Pergamon Press.)

statistical. Thus the probability of having a vanadium surrounded by 0 vacancies decreases with increasing carbon content more rapidly than if distribution is only statistical; similarly, vanadium surrounded by three vacancies is less numerous than is given in statistics. Therefore energy of formation of different sites depends on the number of vacancies, which means that there is interaction between vacancies. This briefly summarized study of Froidevaux and Rossier [1967] clearly shows the existence and the influence of defect interactions which is the first step in short-range ordering.

References p. 126

Another example is offered by nonstoichiometric ferrous oxide, FeO, studied by Roth [1960]. The lattice of the sodium chloride type is made of iron cations octahedrally surrounded by oxygen anions, which build a cubic, close-packed framework. In a stoichiometric oxide, every octahedron contains a ferrous ion; in a nonstoichiometric oxide, some octahedra are empty and two ferric ions are created per vacancy. But, at a second stage, some ferric ions may jump into an interstitial tetrahedral position. For instance, in an

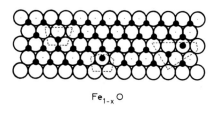

$$Fe_{1-x}O$$

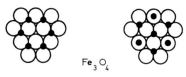

$$Fe_3O_4$$

Fig. 2.26. Oxygen atoms of the (111) plane of iron oxide structure are drawn as large white circles. Some ferric ions have jumped into tetrahedral positions. Associated with two vacancies, they can be regarded as Fe_3O_4 clusters, of which layers are shown in the lower part. (By courtesy of Roth and of Acta Crystallographica.)

oxide of formula $Fe_{0.926}O$, electroneutrality is obtained by the presence of 0.148 ferric ions, of which 0.66 are in tetrahedral holes. Of course, two ions then become close neighbours, one being in a tetrahedral position and the other on an octahedral site. A strong coulombic repulsion should push trivalent interstitial ions back to octahedral sites, unless one assumes that vacancies are closely and directly associated with this interstitial cation (Fig. 2.26). Then electrostatic repulsion is considerably minimized. Such a distribution is characteristic of Fe_3O_4 (magnetite). Since the average size of magnetite domains is small, approximately 8Å, it seems more significant to describe them as regions of clusters due to short-range ordering.

Koch and Cohen [1969], following the work of Manenc and al. [1963, 1964] describing superlattice lines in FeO, recently published a paper throwing new light on the question of short range ordering in this oxide.

These authors interpret experimental X-ray data, collected from a single crystal, as given by clusters which are not magnetite, although they look like in some way, since the position of iron cations in interstitial positions is confirmed. Defects are gathered into clusters for all compositions. They are regularly distributed only at low temperature, becoming non periodically arrayed at high temperature. Between clusters, one finds regular FeO lattice on a length which depends on the average composition. Clusters consist of 13 vacant octahedral sites and 8 tetrahedral sites, which are partly of fully occupied by iron cations. The ions, anions as well cations, surrounding a vacancy complex are slightly displaced, probably because of electrostatic interaction due to a negative balance charge in the cluster.

A third example, the experimental proof of which is less obvious, is found in titanium monoxide. The short-range ordering has been discussed by Arija and Popov [1962]. They have considered that two types of titanium atoms exist, related respectively to the $+1$ and $+3$-states of oxydation. From this it follows that these different titanium atoms cluster in respective small domains, both being crystallized in sodium chloride type, but having different formulae, Ti_2O and Ti_2O_3. In Ti_2O there are oxygen vacancies, and in Ti_2O_3 titanium vacancies; their proportion is such that the general formula is TiO_{1+x}. Thus apparently stoichiometric TiO contains vacancies of both types in equal number, and the systematically found difference between computed and observed values of densities can be explained by this view. Arija and Popov have even tried to evaluate the size of these sub-microdomains, assuming that nonstoichiometry occurs within domains of both kinds, which are bound by surfaces keeping a perfect order. Thus the size of domains must change with deviation from stoichiometry, that is with the change of the ratio O/Ti. They have concluded that TiO_2 domains extend over nine unit cells for $TiO_{0.88}$ and over three unit cells for $TiO_{1.20}$. Change in stoichiometry occurs, under these conditions, as a dynamic process. It is worth pointing out that observed metallic conductivity of this oxide seems in conflict with the coexistence of two states of oxidation. Let us recall that such an assumption had been already put forward by Straumanis and Li, assuming that titanium atoms were at 0 and $+3$-states. Their proposal was based upon X-ray intensity measurements, that is on comparison between observed and computed values, using different atomic scattering factors for titanium according to state of oxidation; however, this seems to be at the limit of reliability and may be open to discussion.

The description of short-range ordering by Arija and Popov starts from considerations which are not structural, and no direct crystallographic study of TiO supports their view. This is why the work of Thornber et al. [1968] is quite interesting. It has been described earlier (p. 108) how zirconium–scandium mixed oxides are made of octahedral blocks. If they are regularly assembled, one gets an ordered compound like those mentioned, $Sc_4Zr_3O_{12}$ and $Sc_2Zr_5O_{13}$. If on the other hand blocks are not regularly ordered although coherently assembled in such proportion that the average chemical formula is verified, a nonstoichiometric compound is constituted.

This mechanism may as well be applied to shear structures. It has been explained that dislocations may generate ordered structures in such a way that a description is an array of regularly repeated dislocation planes. Now, if these planes are assumed first as not regularly spaced and second as distributed in several directions, one gets a disordered system. Vacancies do not eliminate in the right number and a partially ordered compound would contain shear planes bounded by dislocation rings, not quite regularly spaced and not quite parallel. Thus nonstoichiometry may be achieved (Anderson and Hyde [1967]). But in this case, an interesting point must be emphasized. If a crystal must contain a given number of point defects to be in chemical equilibrium at a given temperature, situation is not the same for a crystal containing dislocations, since a dislocation does not correspond to a state of thermodynamic equilibrium. Thus, if nonstoichiometry is due to partial ordering with rings of dislocations, this particular crystal cannot be accepted as being in equilibrium. In that case, a large observed deviation would not correctly be described as nonstoichiometry: this may be experimentally shown when techniques have enough sensitivity to allow the equilibrium to be checked, as for $W_{18}O_{49}$ (Gerdanian and Marucco [1966] p. 90).

The question of thermodynamic equilibrium of nonstoichiometric compounds understood as a microdispersion of a clustered phase in a matrix of a different chemical composition is not so clear. This problem will be discussed in this book by Anderson (Ch. 1). However intergrowed clusters may provide nonstoichiometry as Allpress, Saunders and Wadsley [1969] have elegantly proved it recently using electron microscopy.

In the WO_3-Nb_2O_5 system we have earlier mentioned (p. 116), long-range ordering made of a sequence of blocks have been reported, as shown by the X-ray study of single crystals. Herefrom, it may be assumed that nonstoichiometry arises from the coexistence of different chemical formula blocks arrayed by intergrowth in a non perfectly ordered array. By electron microscopy, an image of the lattice of a single crystal may be observed as a sequence of regularly spaced fringes. From the spacing, because crystalline structures of oxides are known, the chemical nature may be deduced. In some cases, experiments show that the regular spacing is interrupted by one or more fringes of a different spacing. This means that, inside a grain of a given composition, an irregularity occurs in the stacking, because some blocks of a different size have coherently grown; blocks of the host lattice and of clusters have a common side. For instance, the complex following case has been observed. $W_5Nb_{16}O_{55}$ corresponds to $MO_{2.6190}$. If a slight change of composition happens, faults are shown by electron microscopy: 125 fringes due to 4×5 blocks of $W_5Nb_{16}O_{55}$, characterized by 18.5Å equidistance, are associated to 4 fringes due to 4×4 blocks of $W_3Nb_{14}O_{44}$, characterized by 15Å equidistance and to 2 fringes due to 4×6 blocks of $W_7Nb_{18}O_{66}$, characterized by 22 Å equidistance. Macroscopically this grain appears like a nonstoichiometric $W_5Nb_{16}O_{55}$ since its average composition is $MO_{2.6187}$. On the submicroscopic block scale, it is a particular mixture of $W_5Nb_{16}O_{55}$-$W_3Nb_{14}O_{44}$-$W_7Nb_{18}O_{66}$ of which blocks are intergrown.

3.4. The influence of temperature on ordering

An ordered compound accepts some point defects, giving rise to non-stoichiometry, in the same way as tin telluride for example, for thermo-dynamic reasons as developed in the first part. But the following question has now to be raised concerning the ordered skeleton of an ordered com-pound. What is its temperature stability? Thermodynamics tell us that it has to disappear with a rise in temperature. A disordered high-temperature state should be observed. Although some compounds stay ordered till their melting point, many order–disorder transitions are known. The case of uranium oxide U_4O_9, which presents a rhombohedral-cubic transition at 65°C, has been already mentionned (p. 105). Similarly, the ordered iron sulfide, Fe_7S_8, becomes a disordered sulphide of nickel-arsenide type when heated to over 220°C.

The effect of temperature on ordering has also been studied for mixed oxides, such as those formed between zirconium oxide and rare-earth ses-quioxides, compounds thoroughly studied by Collongues and co-workers (Perez y Jorba [1962]). Composition $Zr_2M_2O_7$ corresponds to an oxygen vacancy per seven oxygen atoms and ordering depends on the size of the rare-earth ion as shown earlier. In the same way as the extent of the non-stoichiometric phase, maximum transformation temperature is closely related to the ionic radius of the rare earth: it increases with it (Perez y Jorba [1962]). Thus, $ZrLa_2O_7$ is ordered at any temperature, $ZrGd_2O_7$ becomes disordered at 1550°C (Fig. 2.27) and $ZrDy_2O_7$ is disordered at any temperature.

The way of going from the ordered to the disordered state may be quite progressive. Okasaki [1959] has shown that iron selenide becomes ordered through three successive stages, each corresponding to a larger unit cell. At last, symmetry is lowered (Fig. 2.28).

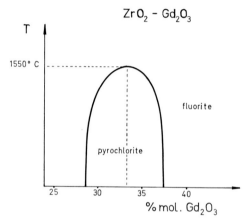

Fig. 2.27. ZrO_2 gives nonstoichiometric mixed oxide of fluorite type with Gd_2O_3. Around composition $Zr_2Gd_2O_7$ the lattice can be of pyrochlorite type because the oxygen vacancies are ordered. Maximum temperature of stability of pyrochlorite structure is 1550°C. (By courtesy of Collongues and of Silicates Industriels.)

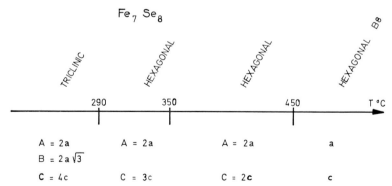

Fig. 2.28. NiAs disordered state of $Fe_{1-7/8}Se$ is stable over 450°C. Under 290°C vacancies are ordered so that formula is Fe_7Se_8 and the lattice becomes triclinic. Between these two temperatures intermediate states of order occur.

Finally let us mention that the degree of ordering has been shown to be dependent on temperature for nonstoichiometric compounds in particular cases, such as FeO. For that purpose, single crystals of FeO have been studied in a X-ray, high-temperature camera. Over 570°C, the limit under which FeO does not exist, the superstructure of FeO appears the sharper and clearer as temperature and composition are related to a point of the iron–oxygen diagram close to the limit of the domain. When temperature grows, superstructure spots become diffuse and their intensities decrease. Thus ordering tends towards short-range ordering (Manenc [1967], Koch and Cohen [1969], cf. pp. 91 and 120).

4. Conclusion

The law of definite proportions has been and still is a good guide for chemists as far as the study of gases and solutions is concerned. Now that experimental methods have enabled an attempt on solid-state problems, regarded for a long time as too difficult, many chemical formulae appeared hard to understand when formulated dualistically. A growing knowledge of the crystal lattices by means of X-ray methods and methods of statistical thermodynamics suggested putting forward the idea of point defects. It first appeared that a solid containing point defects is in a more stable state than one containing no defects, and second that nonstoichiometry is a very general process for nonmolecular crystals. Then experiments revealed that numerous compounds showed deviations from stoichiometry, some of them very large, and the nature of defects was determined.

However, when theoretically computed, thermodynamic state functions were compared to the experimentally determined ones, it rapidly came out that there are seldom cases where defects, which can be regarded as chemical species, are dissolved ideally in the lattice matrix. Considered together with the ordering phenomenon already known for alloys, that was the starting point of the study of order in nonstoichiometric compounds. Ordered compounds were prepared and described cristallographically: examples were found of compounds with ordered anionic or cationic vacancies, or with ordered interstitial atoms. These results are not in conflict with the idea of nonstoichiometry arising from thermodynamic considerations already mentioned. Only the extent of the nonstoichiometry field is greatly reduced since phases described earlier as large were resolved into discrete compounds. However, it has been shown that a high temperature allows of the recovery

References p. 126

of a disordered state of defects. It is therefore, a matter of competition between thermal energy and ordering energy, that is, interaction energy between defects.

Between completely ordered, long-range ordered state and completely disordered state, intermediate states occur in which short-range order is involved. Nonstoichiometry may then be understood as the coexistence of short-range ordered domains of different compositions between them misorientation is very low. Sometimes vacancies also eliminate themselves, yielding sheared structures. This may be another way of interpreting nonstoichiometry on the basis of irregular shearing; again the lattice appears to be made of short-range ordered blocks. This assumption of nonstoichiometry as arising from short-range ordering is actually favoured; examples of such compounds have been described and investigations are presently being developed in this direction simultaneously with the present one of ordered compounds in which ordered units may be distinguished and therefore used for describing nonstoichiometry.

References

ALLPRESS, J. G., J. V. SAUNDERS and A. D. WADSLEY, 1969, Acta Cryst. **B25**, 1156.

ANDERSON, J. S., 1946, Proc. Roy. Soc. **A185**, 69.

ANDERSON, J. S. and B. G. HYDE, 1967, Phys. Chem. Solids **28**, 1393.

ANDERSSON, S., 1967, Bull. Soc. Minér. Crist. **90**, 522.

ANDERSSON, S., B. COLLEN, U. KUYLENSTIERNA and A. MAGNELI, 1957, Acta Chem. Scand. **11**, 1641.

ANDERSSON, S., W. G. MUMME and A. D. WADSLEY, 1966, Acta Cryst. **21**, 802.

ANDERSSON, S. and A. D. WADSLEY, 1962, Acta Cryst. **15**, 194.

ARIYA, S. M. and Y. G. POPOV, 1962, J. Gen. Chem. USSR **32**, 2077.

ASHBEE, K. H. G. and R. E. SMALLMAN, 1963, Proc. Roy. Soc. **A274**, 195.

ATLAS, L. M., 1968, Phys. Chem. Solids **29**, 91.

ATLAS, L., 1969, personal communication.

AUBRY, J. and F. MARION, 1955, C. R. Acad. Sci. **240**, 1770.

BAENZIGER, N. C., H. A. EICK, H. S. SCHULDT and L. EYRING, 1961, J. Am. Chem. Soc. **83**, 2219.

BARTRAM, S. F., 1958, Thesis Rutgers Univ. USA, Dissert. Abst. **19**, 1216.

BARTRAM, S. F., 1966, Inorg. Chem. **5**, 749.

BERTAUT, E. F., 1953, Acta Cryst. **6**, 557.

BEVAN, D. J. M. and J. KORDIS, 1964, J. Inorg. Nucl. Chem. **26**, 1509.

BELBEOCH, B., J. C. BOIVINEAU and P. PERIO, 1967, Phys. Chem. Solids **28**, 1267.

BJERKELUND, E. and A. KJEKSHUS, 1967, Acta Chem. Scand. **21**, 513.

BOKII, G. B. and R. F. KLEBTSOVA, 1965, J. Strukt. Chim. USSR **6**, 866.

BREBRICK, R. F., 1963, Phys. Chem. Solids **24**, 27.

BREBRICK, R. F. and A. J. STRAUSS, 1964, J. Chem. Phys. **41**, 197.

BREBRICK, R. F., 1967, Non-Stoichiometry in Binary Semiconductor Compounds, $M_{\frac{1}{2}-\delta}N_{\frac{1}{2}+\delta}$(c), in: Progress in Solid State Chemistry, Vol. 3, ed. H. Reiss (Pergamon Press) pp. 213–264.

BREBRICK, R. F., 1969, International Conference on Non-Stoichiometry, Tempe, USA.

BROWN, B. E. and D. J. BEERNTSEN, 1965, Acta Cryst. **18**, 31.

CARNAHAN, R. D. and J. O. BRITTAIN, 1963, J. Appl. Phys. **34**, 3095.

CARTER, R. E. and W. L. ROTH, 1967, Symposium of the Nuffield Research Group (Imperial College, London).

CHESTER, P. F., 1961, J. Appl. Phys. Suppl. **10**, 2233.

CHEVRETON, M., 1964, Thesis, University of Lyon, France.

CHEVRETON, M. and S. BRUNIE, 1964, Bull. Soc. Fr. Minér. Crist. **87**, 277.

CONDIT, R. H., 1959, Kinetics of High Temperature Processes (Wiley) p. 97.

EHRLICH, P., 1939, Z. Elektroch. **45**, 362.

EIKUM, A. and R. E. SMALLMAN, 1965, Phil. Mag. **11**, 627.

ELLIOTT, G. R. B. and J. F. LEMONS, 1960, J. Phys. Chem. **64**, 137.

ENGELL, H., 1957, Arch. Eisenhüttenw. **28**, 109.

ERD, R. C., H. T. EVANS and D. H. RICHTER, 1957, Am. Min. **42**, 309.

FLINK, E., G. A. WIEGERS and F. JELLINEK, 1966, Rec. Trav. Chim. Pays Bas **85**, 869.

FØRLAND, K., 1964, Acta Chem. Scand. **18**, 1267.

FROIDEVAUX, C. and D. ROSSIER, 1967, Phys. Chem. Solids **28**, 1197.

FUJIMOTO, M. and Y. SATO, 1966, Japan. Appl. Phys. **52**, 128.

GERDANIAN, P. and J. F. MARUCCO, 1966, C. R. Acad. Sci. **262**, 1037.

GLEIZES, A. and Y. JEANNIN, 1969, J. Solid State Chem., to be published.

GLEMSER, O. and H. SAUER, 1943, Z. Anorg. Allg. Chem. **252**, 144.

GRØNVOLD, F., O. HAGBERG and H. HARALDSEN, 1958, Acta Chem. Scand. **12**, 971.

GRØNVOLD, F., H. HARALDSEN and A. KJEKSHUS, 1960, Acta Chem. Scand. **14**, 1879.

HAGENMULLER, P., J. GALY, M. POUCHARD and A. CASALOT, 1966, Mat. Res. Bull. **1**, 95.

HÄGG, G., 1933, Trans. AIME **105**, 287.

HAHN, H. and B. HARDER, 1956, Z. Anorg. Allgem. Chem. **288**, 241.

HAHN, H., B. HARDER, U. MUTSCHKE and P. NESS, 1957, Z. Anorg. Allgem. Chem. **292**, 82.

HAHN, H. and F. JELLINEK, 1962, Naturwissensch. **49**, 103.

HAHN, H. and P. NESS, 1959, Z. Anorg. Allgem. Chem. **302**, 136.

HANSEN, M., 1958, Constitution of Binary Alloys (Mac Graw-Hill).

HARDY, A., J. GALY, A. CASALOT and P. HAGENMULLER, 1965, Bull. Soc. Chem. **11**, 1056.

HERZOG, E., 1959, Corrosion et Anticorrosion **7**, 281.

HIMMEL, L., R. F., MEHL and C. E. BIRCHENALL, 1953, Trans. Am. Inst. Min. 827.

HUBER, M. and G. LINE, 1953, C. R. Acad. Sci. **256**, 3118.

HUISMAN, R. and F. JELLINEK, 1969, J. Less Com. Met. **17**, 111.

HYDE, B. G., D. J. M. BEVAN and L. EYRING, 1965, Intern. Conf. Electron Diffraction and Crystal Defects (Melbourne) II, Ch. 4.

HYDE, B. G. and L. EYRING, 1965, Rare Earth Res. (Gordon & Breach) **3**, 623.

JACQUIN, Y. and M. HUBER, 1966, C. R. Acad. Sci. **262**, 1059.

JACQUIN, Y. and Y. JEANNIN, 1963, C. R. Acad. Sci. **256**, 5362.

JEANNIN, Y. and J. BENARD, 1963, Advan. Chem. Series **39**, 191.

JELLINEK, F., 1957, Acta Cryst. **10**, 620.

JELLINEK, F., 1962, J. Less Common Met. **4**, 9.

JETTE, E. R. and F. FOOTE, 1933, J. Chem. Phys. **1**, 29.

KAMIGAICHI, T., 1952, J. Sc. Hiroshima Univ. **A16**, 325.

KATSURA, T., B. IWASAKI and S. KIMURA, 1967, J. Chem. Phys. **47**, 4559.

KJEKSHUS, A., 1966, Acta Chem. Scand. **20**, 577.

KJEKSHUS, A. and W. B. PEARSON, 1964, Phases with the Nickel Arsenide and Closely-Related Structures, in: Progress in Solid State Chemistry, Vol. 1, ed. H. Reiss (Pergamon Press) pp. 83–174.

KOCH, F. and J. B. COHEN, 1969, Acta Cryst. **B25**, 275.

LAFFITTE, M., 1959, Bull. Soc. Chim. 1223.

LE BLANC, A., J. ROUXEL and M. DANOT, 1968, Meet. Soc. Chim. Fr.

LIBOWITZ, G. G., 1965, Nonstoichiometry in Chemical Compounds, in: Progress in Solid State Chemistry, Vol. 2 (Pergamon Press) p. 216.

LIBOWITZ, G G. and T. R. GIBB, 1957, J. Phys. Chem. **61**, 793.

LIBOWITZ, G.G. and J. B. LICHTSTONE, 1967, Phys. Chem. Solids **28**, 1145.

McTAGGART, F. K. and A. D. WADSLEY, 1958, Austr. J. Chem. **11**, 445.

MAGNELI, A., 1949, Arkiv. Kemi **1**, 223.

MANENC, J., 1963, J. Phys. Rad. **24**, 447.

MANENC, J. and T. HERAI, 1963, C. R. Acad. Sci. **256**, 684.

MANENC, J., J. BOURGEOT and J. BENARD, 1963, C. R. Acad. Sc. **256**, 931.

MANENC, J., T. HERAI, B. THOMAS and J. BENARD, 1964, C. R. Acad. Sc. **258**, 4528.

MANENC, J., G. VAGNARD and J. BENARD, 1962, C. R. Acad. Sci. **254**, 1777.

MARION, F. and M. CHOAIN-MAURIN, 1962, Chimie et Industrie **88**, 483.

MERING, J. and A. LEVIALDI, 1941, C. R. Acad. Sci. **213**, 798.

MEYER, F. H., O. L. RIGGS, R. L. MACGLASSON and J. P. SUDBURY, 1958, Corrosion, USA **14**, 109t.

MILLER, G. R. and CHE-YU LI, 1965, Phys. Chem. Solids **26**, 173.

MOOSER, E. and W. B. PEARSON, 1959, Acta Cryst. **12**, 1015.

MOROZOWA, M. and L. GUETZKINA, 1959, Vestnik Lenin. Univ. **4**, 128.

OKAZAKI, A., 1959, J. Phys. Soc. Japan **14**, 112.

OKAZAKI, A., 1961, J. Phys. Soc. Japan **16**, 1162.

PEREZ Y JORBA, M., 1962, Ann. Ch. **7**, 479.

PEREZ Y JORBA, M., R. COLLONGUES and J. LEFEVRE, 1959, C. R. Acad. Sci. **249**, 1237.

PEREZ Y JORBA, M., M. FAYARD and R. COLLONGUES, 1962, Bull. Soc. Chim. 155.

POUCHARD, M., A. CASALOT, J. GALY and P. HAGENMULLER, 1967, Bull. Soc. Chim. **11**, 4343.

POUCHARD, M. and P. HAGENMULLER, 1967, Mat. Res. Bull. **2**, 799.

RAU, H., 1967, Phys. Chem. Solids **28**, 903.

RAYNOR, G. V., 1959, Symposium National Pysical Laboratory (Teddington) 3A.

REES, A. L. G., 1954, Trans. Far. Soc. **50**, 335.

REID, A. F., W. G. MUMME and A. D. WADSLEY, 1968, Acta Cryst. **B24**, 1228.

ROTH, W. L., 1960, Acta Cryst. **13**, 140.

ROTH, R. S. and A. D. WADSLEY, 1965, Acta Cryst. **19**, 26.

ROTH, R. S. and A. D. WADSLEY, 1965, Acta Cryst. **19**, 32.

ROTH, R. S. and A. D. WADSLEY, 1965, Acta Cryst. **19**, 38.

ROTH, R. S. and A. D. WADSLEY, 1965, Acta Cryst. **19**, 42.

RÜDORFF, W., 1965, Chimia **19**, 489.

RUYSINK, A., F. KADIJK, A. J. WAGNER and F. JELLINEK, 1968, Acta Cryst. **B24**, 1614.

SAWYER, J. O., B. G. HYDE and L. EYRING, 1965, Bull. Soc. Chim. 1190.

SELTE, K. and A. KJEKSHUS, 1964, Acta Cryst. **17**, 1568.

STRAUMANIS, M. E. and H. W. LI, 1960, Z. Anorg. Allgem. Chem. **305**, 143.

STRAUMANIS, M. E., T. EJIMA and W. J. JAMES, 1961, Acta Cryst. **14**, 493.

THORNBER, M. R., D. J. M. BEVAN and J. GRAHAM, 1968, Acta Cryst. **B24**, 1183.

VAN GOOL, W., 1966, J. Mat. Sci. **1**, 261.

VAN LANDUYT, J., G. GEVERS and S. AMELINCKX 1964, Phys. State Solids **7**, 307.

VOELTZEL, J. and J. MANENC, 1966, C. R. Acad. Sci. **262**, 747.

WADSLEY, A. D., 1955, Rev. Pur. Appl. Chem. **5**, 165.

WADSLEY, A. D., 1955, Acta Cryst. **8**, 695.

WADSLEY, A. D., 1957, Acta Cryst. **10**, 715.

WADSLEY, A. D., 1964, Non-Stoichiometric Compounds (Academic Press) p. 98.

WADSLEY, A. D. and W. G. MUMME, 1968, Acta Cryst. **B24**, 392.

WAGNER, C. and G. LORENZ, 1957, J. Chem. Phys. **26**, 1607.

WAGNER, C. and W. SCHOTTKY, 1931, Z. Physik. Chem. **B11**, 163.

WAGNER, C. and J. B. WAGNER, 1957, J. Chem. Phys. **26**, 1602.

WEHEFRITZ, V., 1960. Z. Physik. Chem. **NF26**, 339.

WESTRUM, E. F., H. G. CARLSON, F. GRØNVOLD and A. KJEKSHUS, 1963, J. Chem. Phys. **35**, 1670.

WESTRUM, E. F., C. CHOU, R. MACHOL and F. GRØNVOLD, 1958, J. Chem. Phys. **28**, 497.

WIEGERS, G. A., 1957, Second Intern. Conf. Solid Compounds of Trans. Met. (Enschede, Netherlands).

ZEMANN, J., 1958, Acta Cryst. **11**, 55.

CHAPTER 3

CRYSTALLOGRAPHIC ASPECTS OF NONSTOICHIOMETRY OF SPINELS

H. JAGODZINSKI

Universitätsinstitut für Kristallographie und Mineralogie, Munich, Germany

1. Introduction

The historical development of Crystallography was in complete agreement with Dalton's law of constant and multiple proportions, although very early chemical studies on alloys showed that there are important deviations from this law. Kurnakow [1914] called compounds, deviating from Dalton's law "Berthollides" in order to distinguish them from the "Daltonides", which are strictly stoichiometric. Without any knowledge of the so-called "averaged structure" of compounds, no clear definition of the "Berthollides" in the solid state can be given, unless the existence of a chemical molecule is responsible for a strict stoichiometric formula. But such a compound does not exist in a strict sense, because growth phenomena and entropy are responsible for deviations from the exact stoichiometry, which cannot be measured with sufficient accuracy in most cases. Since the development of experimental techniques, growing crystals of high purity has been very successful in past years, it is very easy to synthesize nearly perfect stoichiometric crystals, if the composition of the ingredients is strictly stoichiometric. But under free conditions with variability of all thermodynamic parameters, including the concentrations of elements of components, all crystals, liquids or glasses are in general "nonstoichiometric".

The composition of a crystal, grown from a nonstoichiometric phase, is clearly a function of temperature. At $T=0°K$, no nonstoichiometric compound exists in normal equilibrium if the elements involved are chemically different. Thus the ideal nonstoichiometric compounds are those with isotopes, but they are not interesting from a chemical or crystallographic point of view, if magnetic interactions of nuclei are excepted.

It cannot be the scope of this paper to give a general review on stoichio-

References p. 177

metric crystals reported in the literature. The reader, interested in chemical or crystallographic details, should be referred to the book by Mandelkorn [1964], where various authors give more or less completely reviewed many classes of nonstoichiometric compounds. In principle all these pathological substances follow very simple rules, defining the type of the nonstoichiometric substance. In this paper we shall restrict ourselves to crystals, although glasses and liquids are the most general examples on nonstoichiometric compounds, because of their structural disorder. In some cases there is no complete miscibility of liquids and glasses, which shows that there are still tendencies to forming stoichiometric compositions, but the possibilities of solving impurity atoms or molecules are much more abundant compared with crystals, on account of the lower degree of structural order which is typical for these kinds of phases.

Crystals with a perfect three-dimensional order should have a perfect stoichiometric composition, but it must be remembered that a really perfect crystal can not exist on account of the unavoidable surface, which involves crystallographic and chemical deviations as well. Thus it may be concluded that nonstoichiometry and crystal disorder are two correlated properties. It will be shown below that this statement is in general true, but a stoichiometric compound does not necessarily imply an ordered crystal; examples are crystals with a statistical distribution of atoms on certain lattice sites.

The following article is not intended to give a comprehensive review of all observations in crystals; it is nearly impossible to cover the large number of experimental results and their interpretations, which are often not very reliable. For a well established judgement on the chemical or crystallographic background of an observed deviation from stoichiometry, a very comprehensive knowledge of the crystal-structure concerned and its various properties is needed. Therefore the author believes that a detailed discussion of a system with very reliable investigations is more instructive than a detailed review of uncorrelated facts. The spinels have been chosen as an example, because none of the known nonstoichiometric structure-types is based on results of similar reliability and generality. Before entering into details, a brief review of the crystallographic background of nonstoichiometry will be given.

2. Ordered crystals

The macroscopic regular crystal forms often lead to the conclusion that crystals are built up by "units", which are periodically repeated by translation (Franckenheim [1856], Bravais [1850]). If all these units were identical, any crystal showing translation symmetry should be strictly stoichiometric if the surface impurities are negligible. Early X-ray diffraction work on crystals showed sharp X-ray reflections. They were generally accepted as proof of a strictly periodic translation symmetry; but investigations of alloys revealed that mixed crystals of alloys and many other compounds may have a geometrically perfect unit cell; in this particular case two or more kinds of atoms occupy certain atomic positions statistically, but the crystal has no translation symmetry in a strict sense.

Density measurements often result in deviations from stoichiometric composition (non-integral number of formula units in the unit cell). They were interpreted by assuming some defects, such as boundaries of mosaic blocks, dislocations, which lead to lower densities than calculated theoretically with the assumption of a stoichiometric compound, defects of this kind are not of interest here, although they are in some cases correlated with nonstoichiometry.

In an ordered structure with n formula-units in the cell, each kind of atom occupies one or more lattice complexes, which define a certain number of crystallographically equivalent lattice positions. Strictly speaking, chemically equivalent atoms are crystallographically different if they occupy different lattice complexes. The difference in the crystal chemical behaviour of chemically equivalent atoms may sometimes be meaningless, but there are often indications of differences which cannot be neglected. Thus ordered crystals are all stoichiometric, but the reverse is not necessarily true, although most of the known stoichiometric compounds crystallize in strictly ordered lattices.

More accurate studies of single crystals revealed that this picture of the crystal is idealized, even if the crystal diffracts according to the dynamic theory of X-ray diffraction. In the last decade it could be shown that ideal crystals and chemical composition are correlated. Very well developed crystals may be obtained by preparing them with very pure ingredients, such as Si, Ge, but it can also be shown that substances containing impurities, may be very perfect crystals. Natural calcite, $CaCO_3$, is often a very perfect crystal in spite of the fact that it has grown from solutions with a very high impurity content and that the lattice itself contains a lot of impurities which do not disturb the averaged geometry of the lattice to any degree. These crystals may show anomalous absorption of X-rays, which is characteristic of very perfect crystals. Statistical considerations by Wagner and Schottky

References p. 177

[1930] have pointed out that any defect in a crystal is correlated with an unfavourable energy Q, the probability P of which is given by

$$P = e^{-Q/kT} \tag{3.1}$$

$k =$ Boltzmann's constant, $T =$ absolute temperature.

Three fundamentally different cases were discussed:

1. substitution   (interchanges of atoms)
2. interstitials   (Frenkel defect)
3. voids   (Schottky defect).

Any crystal shows a small range of miscibility if the Q's – under the conditions in question – are large compared with kT. If there are no strictly equivalent atoms, the only stoichiometric crystals should exist at $T = 0$; with increasing temperature T an exponentially increasing number of defects of different types has to be expected, which normally are randomly distributed over the lattice sites. In crystals with a certain concentration of defects, correlations between them are no longer negligible and the problem of their distribution becomes extremely difficult (cooperative phenomena). At very high temperatures certain chemically related elements often form a complete series of mixed crystals in which any composition may be realized. Nonstoichiometric compounds with a statistical distribution of atoms on one or more symmetrically equivalent atom positions result. Examples are the alloys, some ionic lattices, etc. Thus the nonstoichiometric crystal may belong to a geometrically perfect lattice; the best examples are single crystals in the silver-gold system, where a complete series of mixed crystals exist. But in most cases the distribution of atoms is not random; they often form clusters or superstructures in small parts (domains) and there are some doubts on the limits between stoichiometry and nonstoichiometry.

It should be pointed out here again that stoichiometric compounds do not necessarily have ordered lattices; in the case of the spinels AB_2C_4 there are two structures. In the so-called "normal structure" A, B, C occupy three symmetrically different lattice complexes in the space group F d3m. On the other hand in the "inverse structure", where the same lattice complexes are occupied, B atoms are on the sites originally occupied by A and the remaining B and A atoms occupy the original B-positions statistically. It is justified therefore to call the inverse spinel structure crystallographically nonstoichiometric, although the compound may be stoichiometric from a chemical point of view.

3. Disordered crystals

3.1. X-ray diffraction

As has been pointed out above, nonstoichiometric compounds should crystallize in lattices which are not strictly periodic; but it is well known that sharp reflections are observed, as long as the lattice remains periodic, at least as regards its geometry; in other words, there should be an averaged structure with strict translation symmetry. The observed sharp reflections may be used for Fourier-synthesis. It is generally known that an ordered lattice causes diffraction peaks, the amplitudes of which are calculated according the equation

$$F(h) = \sum_v f_v \exp(2\pi i h r_v) \qquad (3.2)$$

with $h = ha^* + kb^* + lc^*$

a^*, b^*, c^* are translation-vectors of the reciprocal lattice and h, k, l indices of the reflecting planes,

and $r_v = x_v a + y_v b + z_v c$ with

a, b, c translation-vectors of the crystal lattice,

x_v, y_v, z_v components of the atom in position v and

f_v = scattering amplitude of the atom on site v.

In the case of disordered lattices with a perfect geometry of the averaged structure the mean structure amplitude determines the intensities of sharp reflections:

$$F(h\,k\,l) = \sum_v \sum_\mu p_{v|\mu} f_v \exp\{2\pi i(hx_\mu + ky_\mu + lz_\mu)\} \qquad (3.3)$$

where $p_{v|\mu}$ is the a priori-probability of finding an atom of kind v on the atomic position μ. Positions belonging to the same lattice complex have equal $p_{v|\mu}$. The "lattice complex" is determined by the symmetry of the averaged structure. If displacements from the lattice site occur, a generalized displacement function can be used to describe the averaged electron density on the site μ which may include displacements due to vibrations. In this case f_v has to be replaced by $f_{v|\mu}$, indicating that the averaged electronic density is not only dependent on the kind of atoms. Equation (3.3) gives the correct solution for small concentrations of defects and for the integrated sharp reflections only. In addition to these sharp reflections, diffuse ones may be observed, which are determined by the distribution functions of the atoms

References p. 177

$p_{ik|v\mu}(r_m)$, which define the a posteriori-probability of finding an atom of kind k on a lattice μ in the unit cell, which is $r_m = m_1 a + m_2 b + m_3 c$ apart from a cell, where site v is occupied by an atom of kind i. These correlation functions, which have to be defined for each lattice complex of the averaged structure, are very complex in case of complicated lattices. They are simple only if a single lattice complex of the averaged structure has to be considered.

Equation (3.3) already contains the long-range order part of the $p_{ik|v\mu}(r_m)$; therefore new functions $m_{ik|v\mu}(r_m)$, defined by

$$m_{ik|v\mu}(r_m) = p_{k|\mu} + p_{ik|v\mu}(r_m) \tag{3.4}$$

have to be introduced. Using equation (3.4) the X-ray intensities of diffuse reflections I_d are given by

$$I_d = N \sum_i \sum_k \sum_v \sum_\mu \sum_m p_{i|v} f_i f_k m_{ik|v\mu}(r_m) \exp\{-2\pi i(hr_m)\} \tag{3.5}$$

where N is the number of cells of the averaged lattice. In eq. (3.5) the summation over m leads to the final result:

$$I_d = N \sum_{i,k} \sum_{v,\mu} p_{i|v} f_i f_k M_{ik|v\mu}(h) \tag{3.6}$$

where the functions $M_{ik|v\mu}(h)$ are the Fourier-transforms of $m_{ik|v\mu}(r)$. In general only simple monotonously decreasing functions have to be discussed, if a possible superstructure formation is described by subdividing the lattice complexes of the averaged structure in terms of the lattice complexes of the corresponding superstructure. The derivation of the equations here described has been given by Jagodzinski [1963]. In order to determine the disorder observed for nonstoichiometric compounds the following procedure is useful.

1) Careful determination of the averaged structure according to eq. (3.3). The experimental technique should be as advanced as is normally demanded for studies of chemical bonds by diffraction methods; diffuse scattering in the neighbourhood of Bragg reflection should carefully be subtracted from the integrated intensities. A high resultion power and the counter-technique is necessary. The averaged structure shows all deviations from the ideal structure within an error of a few percent, if the intensity measurements were sufficiently accurate. Interstitial atoms are better detected than substitutions or vacancies, if the "termination effect" of the Fourier-synthesis is taken into account.

2) The experimental study of the diffuse background makes possible the comparison with theory, according to eq. (3.5) or eq. (3.6). Here strictly monochromatic X-rays have to be used; film methods are sometimes

advantageous, because of the very low intensity of the diffuse scattering (a big angular area of the diffuse background will be received with a single exposure). The inelastic diffuse background causes much trouble and can only be avoided by the use of a monochromatizer before the counter. In the case of film techniques inelastic scattering can not be suppressed without considerable loss of intensity.

If there is a random distribution of defects, all $m_{ik|v\mu}(r_m)$ are zero unless $r_m = 0$, the Fourier-transforms $M_{ik|v\mu}(h)$ will be constant in this case. A continuously diffuse background results according to eq. (3.6). Even with a strict monochromatic technique with counters and monochromatization of the diffractal intensities, the possibility of detecting the diffuse scattering is very restricted.

No quantitative experiments can be reported so far, but it may be assumed that a few percent of voids may be detected, if all other distortions of the lattice do not cause a more intensely diffuse background. The best result will be obtained by studying the diffuse background at low diffraction angles, where Compton scattering vanishes and lattice distortions make a small contribution to diffuse scattering. The greatest difficulty in detecting defects is to be expected in cases of nonstoichiometric substitution, where the difference in scattering amplitudes of substitutes is the limiting factor of diffuse scattering. In most cases the diffuse intensities due to the strains which accompany all statistical disorder, are much larger than the general diffuse background. Experimental X-ray studies of the system $BaSO_4$–$SrSO_4$ (unpublished investigation of the author's laboratory) revealed that the defects other than the substitution of Ba by Sr make the most important contribution to diffuse scattering. In case of short-range order the distribu-

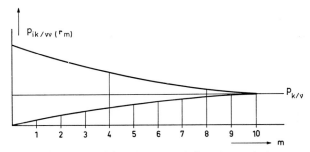

Fig. 3.1. Qualitative behaviour of functions $p_{ik|vv}(r_m)$ and $m_{ik|vv}(r_m)$; the more general functions for different sites are similar, a displacement of positions $r_{v\mu}$ has to be taken into account.

References p. 177

tion functions $p_{ik|v\mu}$ (r_m) are completely determined by the interaction energies between the atoms involved. We shall refer to the possible evalua-tion at the end of this section. An qualitative behaviour of the functions in question is given in Fig. 3.1. for a function, referring to the same lattice complex v; for different lattice complexes very similar functions have to be expected.

According to eq. (3.4) the $m_{ik|v\mu}$ (r_m) are obtained from the $p_{ik|v\mu}$ (r_m) by subtracting $p_{k|\mu}$. This means that these functions tend to zero with in-creasing r_m (compare Fig. 3.1). The explicit evaluation of eq. (3.6) shows that the diffraction pattern is described by diffuse reflections, which are very near the positions of the sharp reflections of the corresponding superstruc-ture. Whether the sharp main reflections, defining the cell of the averaged structure, are accompanied by diffuse ones, is determined by the symmetry relations of the different kinds of atoms involved in the statistics of the lattice complex of the averaged structure. In the case of full symmetry relations for all the kinds of atoms concerned, the sharp reflections are not accompanied by diffuse ones if simple superstructure formation takes place; in all other cases they are.

The limit of detectability of defects increases with decreasing short-range order. The reason is that the integrated diffuse intensity, which is independent of the degree of short-range order, is not concentrated into peaks, which may be detected more easily. With increasing short-range order long-range order may sometimes occur. Then the diffuse reflections, dis-cussed in the case of short-range order, are given by the degree of long-range order as defined by various authors, e.g. Bragg and Williams [1934]. The relationship between short-range and long-range order is rather complicated and cannot be dealt with here. It is completely determined by the geometry of lattice and the interaction energies between the atoms concerned. The formal treatment of long-range order does not involve any difficulty for the diffraction problem; by measurement of the integrated intensities of the sharp superstructure reflections the degree of long-range order may be determined experimentally according to a method given by McGillary and Strijk [1946].

It should be pointed out here that we can get the ordered structure by Fourier-analysis of integrated sharp and diffuse reflections which ac-company the sharp ones. This method assumes that the diffuse intensities may be attributed to a single reciprocal lattice point uniquely.

Great care has to be taken in the case of twinning, which often occurs on account of the higher symmetry of the averaged structure. The Fourier-transformation of diffuse intensities alone can also be carried out, using a sublattice in reciprocal space, which can be equalized with the resolution power of the diffraction apparatus used. The Patterson function of the disordered crystal minus the "peaks" of the Patterson of the averaged structure is received in this way. The short-range order peaks are obtained, negative peaks correspond to lower density for short-range order, when compared with the averaged structure. If the averaged structure has been solved in this way, the interpretation of the maxima, received by Fourier-transformation of diffuse intensities, is not very difficult, unless the structure concerned is not too complicated. Although we have assumed above that only substitutions occur, the more general problem that a substituted atom is accompanied by strains of the lattice does not involve too many difficulties. For the averaged structure an "artificial" temperature factor occurs as has already been shown above. The same is true for short-range order functions $p_{ik|v\mu}$ (r_m), which have to be replaced by functions also defined outside the accurate atomic positions. Let us assume that the replacement of an atom is accompanied by a shorter distance for the immediate neighbours, depending on the statistics of occupation of these sites. Instead of the fixed distance we will have a distribution of distances which are different for the various kinds of atoms involved. The influence on the diffraction pattern is a change of intensities, especially at large diffraction angles, and an increase in diffuseness of short-range order maxima. Strains may often destroy the long-range order of the averaged structure, the originally sharp reflections will become diffuse, too, in this case. A more comprehensive review of this influence will be given below.

Huang [1947] calculated the influence of elastic strains on the diffuse scattering in the case of simple structures. As has been discussed, each interstitial atom or vacancy causes displacements of its immediate neighbours; if these displacements are known, the strains induced around the defect can be calculated by applying the classical theory of elasticity; we shall refer to this contribution later in order to estimate the strain energies involved. Huang calculated the influence on the X-ray intensities. The strain field causes Huang "tails" in the neighbourhood of Bragg-reflections, which indicate whether the strain field was expanding or contracting. Furthermore a small change of lattice constants should be observed. In general the energy of the elastic strain field has an important influence on the activation energy used in eq. (3.1).

If a strong electrical charge is correlated with a defect, it has to be assumed that there are other kinds of defects of opposite charges which interact strongly with each other. In this way groups of correlated defects have to be considered, which cannot be conveniently treated by applying eq. (3.5) or eq. (3.6), although these equations are of general validity. In the case of small concentrations, a different model of the averaged lattice and the group of defects has to be chosen, which will be demonstrated below; the calculation of diffuse scattering will be given in the special case of the spinel structures.

In principle all defects associated with nonstoichiometry have been discussed so far, but sometimes more complicated groups of defects may occur, which make more specialized formulas for X-ray intensities useful. There are some examples of nonstoichiometric compounds whose occupied interstitials form complete chains, with either a rational or an irrational length of the lattice constants. In this particular case one dimension of the crystal remains strictly ordered and the disorder is restricted to the dimensions deviating from the chain direction. The X-ray diffraction pattern (rotation or oscillation photography with the incident X-ray beam vertical to the chain direction) shows diffuse layer lines, the distances between which determine the lattice constant within the included chain. X-ray diffraction pictures of this kind have been published by Lenné [1961], who investigated inclusion compounds of urea. Some X-ray pictures are completely diffuse within the layer, some others show more or less pronounced maxima, which may be interpreted by assuming ordering forces between the chains, either by direct interaction or by indirect correlations over the framework of the matrix. The order problem is two-dimensional, but it should be remembered that it is difficult to get equilibrium conditions. Ordering processes are of the same kind as in the three-dimensional cases, critical temperatures may occur, but relaxation effects make the problem more difficult.

Crystals with two-dimensionally extended inclusions are also observed, but in this case the crystal is no longer coherent, because of the two dimensional net, which "cuts" the crystal into two pieces. This kind of nonstoichiometry may occur, if the sheets of the matrix crystal are either electrically neutral and bound by van der Waals forces only, or ionic (or metallic) bonds are the predominant interaction between matrix and included layer.

There are many examples of this type of nonstoichiometry. The mineral Montmorillonite may take up a variable amount of water (Weiss et al. [1958]); a continuous increase in lattice constants with increasing water content indicates a statistical arrangement of the silicate and the water layers.

Similar observations have been made with the chlorites, where a statistical arrangement of $Mg(OH)_2$ and silicate-layers has been observed (Robinson et al. [1949]). But probably there are electrostatic charge interactions between both types of layers. The statistical problem may formally be treated as one-dimensional, but this is not correct because the crystal represents a three-dimensionally ordered system. Thus this kind of disorder should not arise if there is an unfavourable energy on the part of a fault with respect to the ordered sequence of layers, unless there are precisely the same probabilities for at least two sequences. For further details the reader is referred to the book by Mandelkorn [1964].

3.2. Energy calculation of nonstoichiometric crystals

The calculation of the energy of nonstoichiometric crystals is closely related to the X-ray diffraction problem if only electrostatic forces are considered. Although the present development of band structure and its energy calculation is a powerful tool for describing the electronic structure in solids, it may be useful to derive the energy calculation in terms of electrostatic interactions. The following arguments may justify the procedure described in this paper:

The interaction energy of any system may be separated into three parts:

a) The self energy of electrons and nuclei which describes the interaction of electronic states and the nuclei; this energy differs considerably from the energy of the free state unless only van der Waal's forces are realized; the averaged kinetic energy of electrons has to be included.

b) Charge–charge interactions of atoms and their electrons, which act as a point charge in case of spherically symmetric electron distributions. In this picture each atom is included in a polyhedron, which may be constructed according to Kasper by putting intersecting planes between neighbouring atoms.

c) Electrostatic interactions of higher moments, which are often negligible if charge–charge interactions occur.

Although this separation of energy is arbitrary as there are some doubts as to what part of energy should be attributed to the self energy, this kind of description is formally correct at least in principle and gives the same results as can be derived from calculations of the band energy. For describing order–disorder phenomena the procedure given here results in a more straightforward understanding of cooperative phenomena.

References p. 177

Electrostatic interactions are a function of the distance between inter-acting particles; therefore the same functions

$$p_{ik|v\mu}(r_m)$$

have to be introduced for energy calculation. The whole potential energy U of the crystal given by

$$U = U_s + \sum_m \sum_{i,k} \sum_{v,\mu} \frac{p_{i|v}\eta_i\eta_k p_{ik|v\mu}(r_m)}{|[r_m - (r_\mu - r_v)]|} \tag{3.7}$$

where U_s = self energy to be calculated by a quantum mechanical treatment η_i, η_k = effective charge of atoms i, k, which are not necessarily integers. Similar expressions may be given for interactions of nonspherically sym-metric arrangements; the charges η_i, η_k have to be replaced by dipole or higher moments which are dependent on their orientation; the denominator in eq. (3.7) has to be changed to $|r_m - (r_\mu - r_v)|^n$, where n is characteristic for the kind of interaction. Eq. (3.7) may well be used for calculating lattice energies of ordered and disordered crystals as well.

The self energy U_s may be very sensitive with respect to configurations. In order to calculate this part of energy one should at least know the arrange-ment of the surrounding atoms; the self energy contribution will be given by $U_{ik}(r)$; in a first approximation the averaged configuration may be chosen in order to have an independent contribution to U_s. Thus the energy may be given by

$$U_s = \sum_{\mu,v} \sum_{i,k} p_{i|v} \langle U_{ik}^{conf} \rangle (r_v - r_\mu) \, p_{ik|v\mu}(0) \tag{3.8}$$

where the summation over μ has to include immediate neighbours only, the argument "zero" of p is used for the same reason.

The electrostatic part U_e of interaction energy U may be calculated according a method described by Jagodzinski and Haefner [1966]:

The $p_{ik|v\mu}(r_m)$ are replaced by $m_{vk|v\mu}(r_m)$ according to eq. (3.4).

This method applies to short-range order and long-range order as well according to the arguments elaborated below.

It may now be shown that the electrostatic energy U_e may be split into two contributions U_a and U_o:

$$U_e = U_a + U_o \tag{3.9}$$

where U_a is the electrostatic contribution to lattice energy for the averaged structure and U_o the contribution of short-range order. For U_a Jagodzinski and Haefner [1966] give the following expression:

$$U_a = \frac{1}{(2\pi)^3 V}\sum_{h,k,l} \frac{|F(h,k,l)|^2}{|h|^2}\,|\varphi(h,k,l,)|^2 - \int_{-\infty}^{+\infty} |\varphi(h,k,l)|^2\,\mathrm{d}h \sum_v q_v^2 \tag{3.10}$$

$v=$ volume of the unit cell

$h = ha^* + kb^* + lc^*$

$\varphi(h,k,l) =$ spherically symmetric charge distribution function (without overlap in order to avoid corrections

$$F(h,k,l) = \sum_v q_v \exp\{2\pi i\,(hx_v + ky_v + lz_v)\}$$

$$q_v = \sum_i p_{iv}\eta_{i|v}$$

$\eta_{i|v} =$ effective charge of atom i on site v, this quantity is not necessarily an integer, in order to include compounds the chemical bonds of which are not entirely ionic.

It may be shown that the energy calculation of the contribution of short-range order to the lattice energy follows the same lines as for a random arrangement of charges given by eq. (3.10). Before describing its evaluation, some general remarks shall be given on the transition from long-range to short-range order:

At $T=0$ there should be long-range order if all chemically different atoms have different mutual interaction energies. Thus all atoms prefer a fixed position and thus completely ordered lattice results where each site v is occupied by an atom of kind i. With increasing temperature T disorder will occur by interchange between other sites or interstitials. There are two possibilities, which should be discussed separately.

1) The two (or more) sites are equivalent in the sense that an interchange of all atoms i on site v by atoms k on site μ leads to a symmetrically equivalent arrangement. In this particular case a critical temperature T_c exists where a breakdown of long-range order occurs because the concentration of atoms i on site v is no longer different from the concentration of atoms k on site v; the same is true for site μ. Four a priori-probabilities become equal: $p_{i|v}=p_{i|\mu}=p_{k|\mu}=p_{k|v}$. Similar relations

exist for $p_{ik|v\mu}(r_m)$. The critical temperature T_c is caused by a degeneracy of the statistical problem, which may better be understood by starting from temperatures above T_c: With decreasing temperature, increasing domains of the superstructure observed at $T=0$ occur, but as an interchange of i and k leads to symmetrically equivalent solutions, which have the same probabilities at temperatures above T_c, the increase of domains is limited in the two or three-dimensional case of disorder because of an increase of the unfavourable energy of the boundary. At the critical temperature T_c, the crystal has to realize one of the two equivalent solutions of long-range order. There may also be more than two solutions in more complicated structures.

2) If the interchange is not symmetrical in the sense described above there cannot be a critical temperature T_c, because either of the lattice sites involved is preferably occupied by one of the species of atoms. Therefore even at very high temperatures there is a small, but definite preference by atoms for some site, which is characteristic for long-range order. Thus with increasing T a gradually increasing disorder is observed without any critical temperature. The problem is degenerated for $T=\infty$. This statement is in contradiction to the early theory by Bragg and Williams [1934]; but it should be pointed out that this different conclusion is due to the assumption that the interchange of two atoms is associated with an unfavourable energy V_o, which is zero for a random distribution of the lattice sites involved. But this can only be true if the two sites are symmetrically equivalent, in all other cases it is never zero, because of the preference of the kind i to the site v! Therefore the theory of Bragg and Williams can only be applied to order–disorder phenomena on symmetrically equivalent sites.

In Fig. 3.1 a qualitative description of functions $p_{ik|v\mu}(r_m)$ and $m_{ik|v\mu}(r_m)$ is given; it is obvious that any function $m_{ik|v\mu}(r_m)$ may be described by the corresponding strictly periodic function m', multiplied by the "shape function" already discussed above:

$$m_{ik|v\mu}(r_m) = m'_{ik|v\mu}(r_m)s_{ik|v\mu}(r_m).$$
(3.11)

The "shape functions" are not all necessarily different from each other, because of the symmetry relations between ordering mentioned above. All functions of symmetrically equivalent atoms on equivalent sites will be identical; therefore a careful examination of the symmetry relations of the statistical problem has to procede the mathematical solution of the problem. Now the energy contribution of short-range order to the lattice energy can be calculated as follows: In case of long-range order the various lattice sites have to be renumbered in a manner that each of the equivalent sites of the averaged structure, which have become non-equivalent sites on account of the long-range order, have to be listed as new sites with new a posteriori-probabilities $p'_{ik|v\mu}(r_m)$, which are equal for all corresponding sites of the averaged structure.

The contribution U_{sr} of short-range order to the lattice energy may now be calculated:

$$U_{sr} = \sum_{v,\mu} \sum_{i,k} \sum_{m} \frac{p_{i|v} m'_{ik|v\mu}(r_m) \cdot s_{ik|v\mu}(r_m)}{|r_m - (r_\mu - r_v)|}. \tag{3.12}$$

The equation may be evaluated directly in all cases where the functions $s_{ik|v\mu}(r_m)$ are sufficiently convergent, so that only a few terms in eq. (3.12) have to be considered. On the other hand the summation of ordered structures may be done very easily by computer programmes, which normally operate in reciprocal space. This method uses these Fourier-transforms of the function m' s, which is given by:

$$m' s = \int \overline{M'(h) S(h)} \exp(-2\pi i hr) dh, \tag{3.13}$$

⌒ is the symbol for convolution integral,

where $M'_{ik|v\mu}(h)$ and $S_{ik|v\mu}(h)$ are the Fourier-transforms of m' and s respectively. Now the sum over m in eq. (3.12) may evaluated

$$U_{sr} = \sum_{v,\mu} \sum_{i,k} p_{i|v} \exp\{2\pi i h(r_\mu - r_v)\} \int \frac{\overline{M'_{ik|v\mu}(h) S_{ik|v\mu}(h)}}{|h|^2} dh - U_{sr}^0; \tag{3.14}$$

U_{sr}^0=self energy incorrectly introduced by the integration. Although eq. (3.14) is not very convenient for calculating lattice energies of disordered crystals because eq. (3.12) is more rapidly convergent in the case of short-range order, it gives a more instructive estimate of the contribution of short-range order to lattice energy. It should be remembered that the integral may be replaced by a sum over the reciprocal lattice points if long-range order only has to be assumed. The main difference between the disordered and the ordered state is the "overlap" of maxima, which is completely determined by the convolution of S and M. In general a rather rapid decrease of $S_{ik|v\mu}(r)$ causes slowly decreasing function $S(r)$ with a considerable overlap. This causes a decrease in energy due to short-range order. The energy of the first few terms in eq. (3.14) has little influence on the energy calculation; the correction for self energy in particular may also be important, but in cases of rapid convergence of $S_{ik|v\mu}(r)$ in crystal space the contribution of "overlap" becomes predominant.

The calculation of energy, given above, is based on the incorrect assumption that the disorder is not accompanied by displacements of the atoms. Strains in the lattice are in most cases due to the "repulsive forces" which have been included in the self energy, the decrease of this energy contribution being very rapid with increasing distance. A fair estimate of their energy may be given by taking into account next nearest neighbours only.

Any single fault differing by geometry of the replaced atom or group of atoms and the matrix causes internal strains, as long as the geometrical conditions at the boundary between "defect" and the replaced ordered part are different. This "misfit" causes displacements of atoms at the boundary of the crystals, which contribute essentially to the self energy of the boundary atoms. As a consequence a strain field is induced, which influences a rather large part of the crystals.

The calculation of the energy of this strain field follows the same lines as the strain field of a dislocation: two parts of the energy have to be taken into account.

1) The energy change of the self energy of the boundary atoms.
   This part can only be evaluated by introducing a definite picture of the structure and the interactions in the boundary.
2) Once the displacements of the boundary are given, the components of the deformation tensor are known and the strain field may be determined applying normal theory of elasticity, e.g. the strain energy is completely determined by the "misfit" and the elastic constants of the crystal of the (matrix) crystal.

As the procedure in calculating the strain field energy described is well known, the mathematical derivation will not be repeated here; let us discuss the result very briefly. If the misfit of the defect is not too large the displacements are smaller than the actual atomic distances of the structure, the strain field makes the most important contribution to the energy of the defects. In all other cases, where the defects cause rather large displacements of atoms or are themselves big compared with the strained volume, the self energy of the defect becomes increasingly important.

The energy can be evaluated with the aid of model structure of the defect. Even in the case of electrostatic interaction, it may be shown that the change of energy induced, is primarily restricted to nearest-neighbour interactions: As has been shown below, the energy of the lattice may now be described by 3 parts.

1) The electrostatic energy of the averaged structure as given by eq. (3.10).
2) The averaged potential of the strain field.
3) The interaction of nearest neighbours.

Part 1) of the energy contribution makes no contribution to the averaged lattice energy if two conditions are fulfilled:

a) The averaged displacement function is spherically symmetric.
b) The displacements of the centres of atoms do not exceed a critical amount for neighbouring atoms, such that the convolution of the displacement-functions does not lead to overlap.

Now the spherical distribution functions of charges may be replaced by δ-functions at the centres, which give no change of energies. A necessary condition for the validity of this procedure is that no electronic overlap of atoms itself is involved by the forces; but it may be assumed that this condition is always valid, the boundary itself excepted, where the interaction energies have to be calculated.

It is very important to note here that the complication of strains does not involve any difficulty, if the displacements of neighbours are not big, even if the complete long-range order is destroyed by the strain-fields in the crystals. This conclusion is very important from a general point of view: The complete statistics of a lattice are not changed in principle by introducing statistics of lattice sites, in the sense that a lattice site is only determined in the immediate neighbourhood of a given atom. Thus the geometrical long-range order of the translation-lattice is not necessarily the result of the cooperative phenomenon. Consequently there need not be a critical point for the transition from the crystalline to the liquid state. The breakdown of lattice-geometry is an indicator, but not necessary for the existence of the critical point. SiO_2 may be one example of such non-critical transition from the crystalline to the "liquid" state. The melting point of crystals is due the sudden increase of entropy, associated with the symmetry degeneracy in a statistical sense. Thus the melting of crystals is a more general order–disorder transition, which is critical in the sense of an order–disorder transition; relaxation may have an important influence on the transition.

Although this conclusion seems to be in contradiction to the experimental fact that "melting" is generally observed in crystals the existence of mixed crystals with missing long-range order of the translation indicates that the geometry of the lattice is not an essential part for the energy and configurational entropy. But it should be pointed out here that the entropy of vibrations has been neglected so far; as entropy plays an important role in this transition, it may well be that the geometrically perfect unit cell is favoured by vibration entropy and our conclusion may be incorrect from this point of view.

The general theory of the disordered crystal may now be outlined as follows:

The number of states associated with any energy of short- or long-range order may be calculated with the aid of statistical calculations of the problem. Thus the entropy S may be determined and

$$F = U - TS \qquad (3.15)$$

can be calculated. The problem of finding the minimum of F is completely determined by energy calculations; because the entropy of any curve is given by the shape functions s. Thus the maximum entropy of s may be found by studying the overlap caused by s in reciprocal space. Although this procedure is rather tedious, the shape functions may be constructed with approximation methods. The change ΔS associated with a change ΔU, gives the complete solution of the problem, the equilibrium curve for S is given by $T\Delta S = \Delta U$. The greatest difficulty is the calculation of correct entropy. An approximate solution may be given, but the procedure cannot be explained here on account of the complexity of the problem.

4. Nonstoichiometric spinels

4.1. Structure of spinels

The great abundance of compounds AB_2C_4 in ternary systems indicates a great stability, which should be due to the favourable geometry of the various structure types of this composition. The chemical bond is predominantly ionic, if complexes AC_4 are agreed upon as being ionic. The structure types are in most cases typical for the sizes of atoms or atomic groups of the compound in question.

The spinel structure may be described as an almost perfect cubic close packed arrangement of oxygens; the A-element occupies a small fraction ($\frac{1}{8}$) of the tetrahedral holes available in close packed structures. The stability of any ionic lattice may be checked by consideration of the distribution of cations, which should be as homogeneous as possible. The distribution of cations in the spinel structure can be described as follows: Half of the octahedral holes are occupied by B-cations, they form chains in the [110] direction. Each (100)-net is composed of alternating occupied and non-occupied chains as shown in Fig. 3.2. If the direction of the occupied chains was the [110] direction, the adjacent equivalent layer has the cation chains

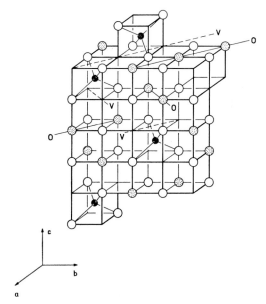

Fig. 3.2. Structure of spinels; open circles = oxygens, full circles = A-cation positions, shaded circles = B-cation positions. Only one layer (100) is shown for the sake of clarity. The remaining layers may be easily constructed applying the following procedure: Occupied (O) and vacant (V) chains of B-cations alternate into the symmetrically equivalent [110]-directions. At the centre of a cross of two vacant chains the A-cation is located. The A-cation pushes the neighbouring oxygens into the corresponding four [111]-directions; in this way the idealized close-packing of anions is distorted.

in the [1$\bar{1}$0]-direction. Where two chains of voids cross there is a high "charge defect" with respect to the rock salt arrangement (completely filled octahedral holes). The holes form a regular tetrahedron in the centre of which the cation A is located. This cation is four-coordinated with respect to anions. Fig. 3.2 shows the structure of such a double net. The two adjacent nets have the same arrangement in principle, but apparently the rows of cations [110] are displaced with respect to the next layer but one. The structure may be described in terms of AO_4-tetrahedra which are connected by the B-cations. The distances AO and BO determine the *a*-lattice constant. Obviously the anions are not ideally close packed, as the distance A-C is in general larger or smaller than the $\frac{1}{8}\sqrt{3}a$, which is the distance A-O for an ideal close packed arrangement of oxygens.

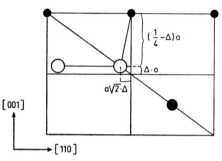

Fig. 3.3. Calculation of important distances of the spinel structure as a function of the parameter *u* of the oxygen lattice complex. Small black circles = B-cations, large black circles = A-cations, open circles = oxygens.

According to Fig. 3.3 the smallest distances A-C (d_A), B-C (d_B) and C-C (d_C) are given by

$$d_{AC} = d_A = a\sqrt{3}(\tfrac{1}{8} + \varDelta) \tag{3.16a}$$

$$d_{BC} = d_B = a\sqrt{\{(\tfrac{1}{4} - \varDelta)^2 + 2\varDelta^2\}} \approx a(\tfrac{1}{4} - \varDelta) \tag{3.16b}$$

$$d_{CC} = d_C = a\sqrt{2}(\tfrac{1}{4} - 2\varDelta). \tag{3.16c}$$

d_A and d_B determine $\varDelta$ and a, applying the following relations

$$\Delta = \frac{2d_A - d_B\sqrt{3}}{8(d_A + d_B\sqrt{3})} \tag{3.17a}$$

$$a = \frac{8}{3}\left(d_B + \frac{d_A}{\sqrt{3}}\right). \tag{3.17b}$$

Equation (3.17b) is valid for small values of Δ. But as Δ_{max}, actually observed in oxide spinels is 1.5×10^{-2}, the correction in eq. (3.16b) is less than 2×10^{-4}, which is negligible remembering that the accuracy d_A and d_B can not be better than 10^{-2}; the known values are most probably much less accurate.

d_C in eq. (3.16c) has a minimum value of closest approach (d_C^0), which fixes a minimum lattice constant,

$$a_{min} = \frac{d_C^0}{\sqrt{2(\tfrac{1}{4} - \Delta)}}$$

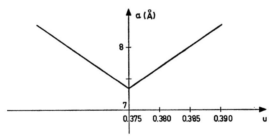

Fig. 3.4. The minimum lattice constant a of oxygen spinels as a function of the parameter $u = \tfrac{3}{8} + \Delta$, caused by oxygen contact.

Fig. 3.4 shows this important relation for oxide spinels. Structures with a possible anion contact may be easily found, $MgAl_2O_4$ is rather close to such a contact of anions.

The absolute minimum will be obtained with $u = 0.375$ where

$$a_{min} = 2\sqrt{2}d_C^0 = 2.83d_C^0 .$$

For oxide-spinels this value becomes $(d_C^0 = 2.6 \text{ Å})$ 7.4 Å, which is in agreement with all experimental observations reported in the numerous papers on oxide spinels. Eqs. (3.17) may also be used to fix a minimum for d_B or d_A, using the ionic radii of the corresponding ions with the given valency.

It should be pointed out here that the valency of A and B may be different, the following table gives the valencies observed so for oxide spinels (Table 3.1).

Table 3.1

Possible cation charges of spinels

Charge of A-cations	Charge of B-cations
2	3
3	2.5
4	2
6	1

Nonstoichiometric spinels, which play an important role in this paper, may have non-integral values; $B = 2.5$ means that B_2 is replaced by $B_1 + B_2$, where B_1 is divalent and B_2 trivalent.

With $d_C = 1.9$ Å the minimum of d_A can be calculated with the aid of eqs. (3.17). This $d_{A\min}$ will be about 1.7 Å. This value limits the range of stability of spinel structures, because bigger values of d_B increase this limit. Thus it may be concluded that r_A (radius of cation A) cannot be smaller than 0.4 Å. This may well be checked by the known structures; the spinels $MnAl_2O_4$–$MgAl_2O_4$–$BeAl_2O_4$ show this fact clearly. In case of $BeAl_2O_4$ Al^{3+} and Be^{2+} could just fill the holes of close-packed oxygens almost perfectly; but this compound crystallizes in another structure type (Olivine structure Mg_2SiO_4). It will be shown below that the non-existence of the spinel $BeAl_2O_4$ is due to the decrease of lattice energy with decreasing u. Thus it may be concluded that the deviation from the close-packed arrangement of anions is essential for 2, 3-spinels.

The so called "normal" structure of spinels, discussed so far, describes just one arrangement of cations. There are some other arrangements, one of them the so called "inverse" spinel structure, where B occupies the tetrahedral holes, while A and B are distributed statistically over the octahedral positions. Thus the crystal chemical formula should be

$B(A,B)_2O_4.$

As A and B generally may have different valencies, it should be expected that A, B are not randomly distributed over the octahedral sites B. At first sight an alternating order of the chains of cations (see Figs. 3.2 and 3.5) into all 6 [110]-directions would be the most stable one. This arrangement is not compatible with an ordered structure, at least four of the six [110] symmetrically equivalent directions show deviations from the idealized arrangement described. Possible superstructures have been explicitly described by Hafner [1960], in principle the superstructures may be easily found with the aid of Fig. 3.5.

The arrangement of the B-cations may be described by introducing

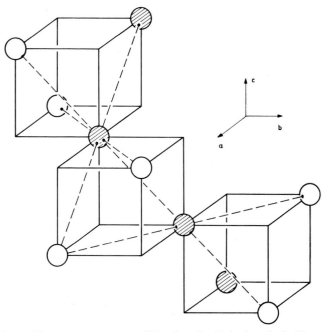

Fig. 3.5. A possible 1:1 superstructure of B-cations B_1, B_2 in the spinels. The superstructures are formed by combining various cubes with different cation arrangements. Open circles: B_1-cations, shaded circles: B_2-cations.

tetrahedra B_4, occupying 4 corners of a cube the centres of which are linked over free corners to a diamond arrangement. Consequently the Al are similarly distributed as the oxygens in high-cristobalite. It may easily be seen that different arrangements are possible for a 1:1 superstructure. Other superstructures have been described for 1:3 and 1:5 ratios of two kinds of ions. Most of the superstructures observed so far are not superstructures of the spinel structure; in a strict sense a change of symmetry takes place, which is often demanded by the change of extinction laws of

X-ray diffraction. Some of the superstructures may be cubic, but others are tetragonal or even of lower symmetry. Submicroscopical twinning may be one of the reasons why the deviation from cubic symmetry is not observed frequently.

For corresponding chemical compositions superstructures on the A-sites are also possible. For different valencies the most probable super-structure is the sphalerite arrangement, which is the most simple super-structure of the diamond arrangement of A-cations. The discussion of possible superstructures on a lattice complex is extremely interesting for nonstoichiometric compounds, because many of these superstructures are correlated with critical temperatures, where the ordered superstructure breaks down and a statistical arrangement can be observed; deviations from stoi-chiometry are very usual in these cases. A change of composition may lead to a decrease in the critical temperature.

4.2. Electrostatic energy and cation distribution

Energy calculations of the electrostatic part of the energy of spinels have been done by Verwey et al. [1948]. The procedure follows the same arguments as given in the preceding chapter on energy calculations in disordered lattices. First the electrostatic energy of a lattice with random distribution of cations is calculated if one of the sites shows disorder. Thus the energy can be given

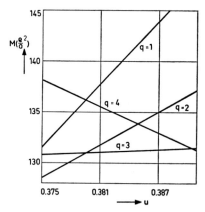

Fig. 3.6. Averaged electrostatic energy of various spinels as a function of q (q = averaged charge on A-sites) and the parameter u. In statistically ordered spinels non-integral values of q are possible.

in terms of q_A, where q_A is the averaged charge on the A-sites. As the number of anion vacancies ever observed is very small, all spinels may be described with the formula

$$(A_x, B_y, \ldots, \square_z)(A_{x'}, B_{y'}, \ldots, \square_{z'})_2 \, O_4.$$

$\square_z$ represents the number of vacancies on the two positions. If the charge of the A-sites (first bracket) is given, the charge on B-sites is completely determined by electrostatic neutrality. Fig. 3.6 shows the results, as given by Verwey et al. [1948]. The electrostatic energy does not vary with the parameter u for $q_A = 3.1$. For smaller values of q_A large parameters are preferred, for larger values of q_A small parameters should occur. Although the repulsive forces are not included, it may be concluded that they make a nearly independent contribution to lattice energy, if the ion may be considered to have the same "hardness". As we are primarily concerned with oxide-spinels, the Born repulsive exponent does not vary for the anions, and the difference of cations are not too big, Li, Mg expected. In case of anion-contacts an additional contribution to repulsive energy has to be introduced, which may change the lattice energy within a few percent.

A more or less important contribution to lattice energies has to be expected for nonstoichiometric spinels as a function of short-range order where the possible long-range order is missing. For long-range order, this calculation has been done very regularly by de Boer et al. [1950], van Santen [1950] and by Goerter [1954]. These values are given on the following table.

Table 3.2

Additional electrostatic energies for superstructures		
A site order 1:1	0.5 (e^2/a)	Δq^2
B site order 1:1	1.0 (e^2/a)	Δq^2
B site order 1:3	0.7 (e^2/a)	Δq^2

Δq is the charge difference between cations participating in the ordering process. In case of 2–3 spinels $\Delta q^2 = 1$ is valid, and the contribution to the lattice energy is very small, but I think that all these estimates given in the literature are incorrect for nonstoichiometric or disorderd spinels, because the strains of the lattice play an important role for energy calculation.

In the preceding section an estimate has been given of the contribution of short-range order to electrostatic energy, this contribution depending largely on the interaction energies (as indicated in Table 3.1) and the repulsive forces. As the ions, ordered statistically, are in general different in size, the repulsive forces between cations and anions will be responsible for a contraction or dilatation of the surrounding oxygens, a displacement from the averaged position will occur. In the ordered structure this contribution may favour or disfavour ordered structures according to the geometry of cation–anion distances in the corresponding ordered structures. The favourable or unfavourable contribution may amount to a few percent of the whole electrostatic energy for a 1:1 order on B-sites or A-sites. It may regulate the repulsion energy of the lattice if both sites are involved in the statistical order (in this case 5% change of repulsion energy for the ordered state is not impossible compared with the completely disordered state and when the size of the ions are very different). As the charge contribution to lattice energy, according Table 3.2, may be very small (less than 1% for charge difference $\Delta q = 1$), this contribution to long-range order may be very important.

Generally the contribution of short-range order may well be estimated with the aid of the critical temperature T_c, where the breakdown of long-range order takes place.

Unfortunately T_c cannot be observed in many cases, because no ordering process can take place without diffusion, which is largely determined by the charge and the size of ions in question. Li and voids may diffuse quite easily, therefore ordering in spinels, containing Li, is very common. Examples are: Fe $[Li_{0.5}Fe_{1.5}]O_4$, $Zn[Ti_{1.5} \square_{0.5}]O_4$, $(Li_{\frac{1}{4}}Fe_{\frac{1}{4}}^{III})G_2O_4$ and others. Another contribution to energy is due to polarization of the anions. It is well known that polarization effect is mainly influenced by the charge distribution: high symmetry, big coordination numbers and small charges restrict this contribution to a minimum. Each anion is almost tetrahedrally surrounded by 4 cations; in the usual normal spinel, say $MgAl_2O_4$, O is surrounded by 3 Al^{3+} and 1 Mg^{2+}, which may be considered as splitting into 2^+ charges arranged at the corners of a nearly regular tetrahedron ($d_A \approx d_B$ in this particular case) and three 1^+-charges arranged in the form of a trigonal pyramid. The latter charge distribution is polar with respect to the centre (position of O) and makes the most important contribution to polarization of the oxygens.

The amount of this additional energy has been estimated by Smit et al.

[1962]. In the case of 2,3 spinels the authors found a contribution of about $8e^2/a$ with $u=0.385$ and nearly zero with $u=0.379$ and the inverse spinel structure; but it should be borne in mind that these calculations are very rough estimates only, which have to be amended considerably by taking short-range order into account.

Another small contribution to lattice energy is the splitting of energy states of d-electrons in the crystal field of the two possible sites of cations. This part of the energy may be important for transition metals, which may have a great number of d-electrons. According to theoretical considerations by Dunitz and Orgel [1957] and McClure [1957] these contributions may be given in terms of site preference for A or B-sites, but the energies involved are very small with respect to all other energy corrections. It seems most unlikely that they have an important influence on cation distribution. Table 3.3 gives a rough indication of this site preference.

Table 3.3

Preference of octahedral sites (in units e^2/a and k cal/mole) for some elements

Cr^{3+}	37.7 k cal/mole	$\sim 1\ e^2/a$
Mn^{3+}	22.8 k cal/mole	$\sim 0.6\ e^2/a$
Ni^{2+}	20.6 k cal/mole	$\sim 0.5\ e^2/a$
Cu^{2+}	15.2 k cal/mole	$\sim 0.4\ e^2/a$
V^{3+}	12.8 k cal/mole	$\sim 0.3\ e^2/a$

From Table 3.3 it may be concluded that for Cr^{3+} only d-electrons may have an important influence on the cation distribution. Nevertheless the compilation of various spinels by Blasse [1964] (pp. 20 ff.) shows that for transition metals this relationship seems to be fulfilled.

The following comment on this question should be added here. The transition between normal structure and inverse structure cannot be correlated with a critical temperature. If there is no preference for either site, the random distribution of cations which corresponds to the composition

$$[A_{\frac{1}{3}}B_{\frac{2}{3}}][A_{\frac{2}{3}}B_{\frac{4}{3}}]O_4$$

should be realized. This crystal chemical formula has been observed for $FeMn_2O_4$, but it should be pointed out that accurate knowledge on cation distribution in spinels is not well established as many observations are based on X-ray powder diffraction data, which give reliable results only if the scattering amplitude of the two or more cations involved have sufficiently different scattering amplitudes. In case of a preference of A (or B) for one of two sites the "preference energy" of that cation and the temperature determine their distribution. There should be a continuous change from an

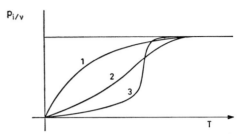

Fig. 3.7. A priori-probabilities $p_{i/v}$ as a function of temperature T for statistics on non-equivalent lattice sites. On the disfavoured site $p_{i|v}$ changes from zero to the value of a random distribution. 2, 3 demonstrate the influence of strains, which may simulate a critical temperature T_c.

ordered structure to random distribution of cations. But the repulsive forces will have a considerable influence on the distribution, which is correlated with the degree of order. The amount of energy necessary for an exchange of an AB-pair of atoms from preferred sites will certainly tend to zero for the repulsive part, because of the most probable relation that the distortions generated in a randomly ordered lattice are negligible compared with the ordered states. The site preference energy for the two symmetrically different sites remains. Thus the cation distribution on A-sites, as a function of temperature, may behave qualitatively as demonstrated in Fig. 3.7. The value of the a priori probability $p_{i|\mu}$ of an atom of kind i on site μ, which is zero at $T=0$ (preference of another site v) is largely determined by the energy difference and short-range order forces. These curves may be estimated by consideration of energy–entropy relations of the cation distribution. Thus it may be concluded that strains may simulate a critical temperature T_c, but it is not really a critical one, if the problem is clearly non-degenerated, which is generally true, if the energy of the site preference is sufficiently large. The cation distribution in normal Mg–Al spinels has been checked by Widman [1966] (unpublished "Diplomarbeit"). It could be shown that synthetic spinels have a different cation distribution at low temperature, which is indicated by changes of lattice constants (Jagodzinski and Saalfeld [1958]) and change of the parameter u. Unfortunately X-rays cannot differentiate between Mg^{2+} and Al^{3+}, because of the same number of electrons and the small difference in radius of the two ions. But this small difference is sufficient to reveal strains in the lattice, which formally correspond to a higher "temperature factor" of the Fourier-synthesis of the crystal. Figs. 3.8a–c show the Fourier-projections of three spinels $MgAl_2O_4$. Fig. 3.8a demonstrates a synthetic spinel $MgAl_2O_4$ with a cation distribution, differing not very much from that of a natural spinel (compare Jagodzinski and Saalfeld [1958]); the "normal picture" of a synthetic spinel is shown in Fig. 3.8b, where considerably lower maxima of electron density are realized. The cation-distribution may be quite different for spinels at low temperatures, where the distribution may be "frozen in". This result is in good agreement with infra-red absorption measurements by Hafner and Laves [1961] and with paramagnetic resonances of Al^{27} in $MgAl_2O_4$ observed by Brun and Hafner [1962]. The first named authors pointed out that heating a spinel up to 900°C results in a broadening of absorption peaks, which corresponds to a considerable distortion of the lattice, but they could not find any change when tempering the crystal at 700°C. This behaviour can be readily understood: diffusion of Mg or Al in the spinels is more difficult in the strictly ordered phase, therefore the breakdown of cation order takes place, as shown on Fig. 3.7, at a well defined temperature, but this temperature is, according to the discussion given above, certainly no critical one, the transition is therefore irreversible and tempering does not cause the rearrangement of the cations of the crystal. Unfortunately the number of high-temperature measurements available is too low. The clear deviation of u as a function of prehistory or temperature from the ideal value of natural spinel $MgAl_2O_4$ shows the direct correlations between cation-distribution and strains (compare Figs. 3.8a–c). Unfortunately these results are only qualitative with respect to the cation-distribution and should be put on a quantitative basis with the aid of neutron-diffraction, where Mg^{2+} and Al^{3+} have sufficiently different scattering lengths. The very careful measurements of Widman with spherical samples of 0.3 mm diameter, applying the counter technique with Mo-radiation, give a good indication that the data, already published by Saalfeld and Jagodzinski [1958], are at least correct in principle.

A great many spinels containing transition metals have magnetic interactions in addition which are of considerable technical importance in the case of the ferrites. Magnetic bonds may have an influence on the cation-distribution and the stoichiometry of the crystals. The order of magnitude of these

References p. 177

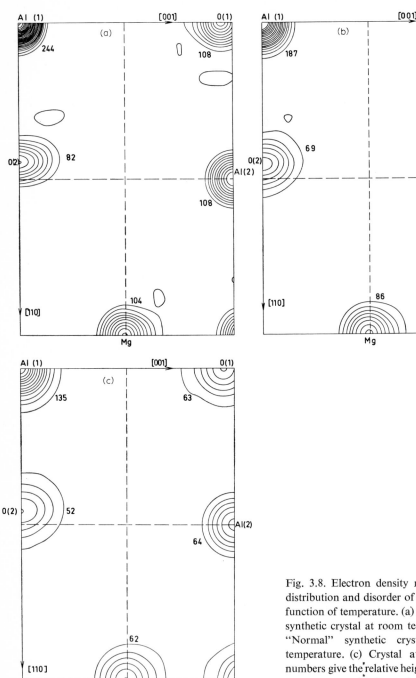

Fig. 3.8. Electron density map of cation distribution and disorder of $MgAl_2O_4$ as a function of temperature. (a) "Exceptional" synthetic crystal at room temperature. (b) "Normal" synthetic crystal at room temperature. (c) Crystal at 850°C. The numbers give the relative heights of electron density peaks.

interactions can well be judged by checking the Curie or Néel-temperature. Table 3.4 gives the Curie-temperatures of a selection of ferrites.

Table 3.4 shows that the magnetic interaction energies are considerable. But this high contribution is only observed with elements, such as Fe, which

Table 3.4

Curie-temperatures of some ferrites

Compound	T_C	Compound	T_C
$NiFe_2O_4$	858°C	$CoFe_2O_4$	795°C
$MgFe_2O_4$	715°C	$Ni_{5/3}(FeSb)_{1/3}O_4$	625°C
$Li_{0.5}Fe_{2.5}O_4$	955°C		

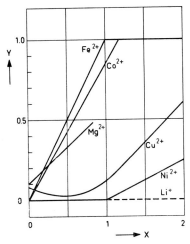

Fig. 3.9. Cation distribution in spinels $Me^{2+}Fe_{2-x}Al_xO_4$. y gives the relative amount of Al on A-sites (after Brasse [1964]).

have extremely high magnetic interactions. On account of the convenient possibility of changing magnetic properties by preparing mixed crystals, nonstoichiometric compounds with the general composition

$$Al^{2+} Me_x^{3+} Fe_{2-x}^{3+} O_4$$

have been studied with respect to their cation distribution and magnetic properties. Fig. 3.9 shows the change of cation distribution of divalent ions in the case of $Me^{2+}Fe_{2-x}O_4$, where Fe^{3+} is replaced by Al^{3+}. In Fig. 3.9 the

References p. 177

amount of divalent ions X on site A is plotted against the parameter $x (x = 0$ means inverse structure, $x = 1$ corresponds to the normal structure). Fig. 3.9 has been taken from Brasse [1964], further results are given in this review article for replacement of Fe^{3+} by Cr, Rh and other elements. In a recent study by Rieck and Diessens [1966], the distribution in Mn, Fe-spinels with two and three-valent ions of both kinds are studied. This system is rather complicated, because of the two valency states of the elements. To judge the result, it should be noted that a charge transfer according to the equilibrium

$$Fe^{3+} + Mn^{2+} \rightleftarrows Fe^{2+} + Mn^{3+} + 0.3 eV$$

shows a tendency towards the left side. An excess of trivalent ions is also possible, but tends to be low. Using the crystal chemical composition formula

$$(Mn_{x-y}Fe_{1-x+y})(Mn_y Fe_{2-y-z} \square_z)$$

the distribution of ions Mn, Fe could be determined as function of x. The number of voids on B-sites, $\square_z$ always remains very small. As an increase of three-valent ions accompanies the increasing number of vacancies, the oxidizing conditions during preparation play an important role, but the concentration of vacancies did not exceed 2% in most cases. As the accuracy of the determination by X-ray intensity measurements should lie in the same order of magnitude, the results are not very reliable. Apparently samples synthesized according to Verneuil's method have a higher excess in oxygen and consequently more cation vacancies. This observation is in good agreement with observations on Mg–Al spinels, discussed below.

Earlier neutron diffraction measurements by Hastings et al. [1956], Alperin [1962], Yamzin et al. [1968], Kleinstück et al. [1965] and other authors did not have the same results, so that a reinvestigation seemed to be desirable.

Since the chemical compositions x for spinels $Mn_x Fe_{1-x} Fe_2 O_4$ are given by mixing the corresponding quantities of MnO and $Fe_2 O_3$ and the oxidation conditions regulated by a mixture of CO and CO_2, the state of oxidation was fairly controlled for the spinels studied by Rieck et al. [1966]. Fig. 3.10 shows the results of their evaluation of X-ray powder diffraction photographs of homogenized samples, for the approximate compositions $x = 0.5$, 1.0 and 1.5. The difference between quenched and slowly cooled samples is not very important. Obviously the cation exchange between octahedral and tetrahedral sites increases slightly for quenched samples, since their cation distribution corresponds formally to a higher temperature. It seems most likely that the equilibrium curve for $T \to 0$ corresponds to a strong preference of Mn for tetrahedral sites (in the presence of Fe).

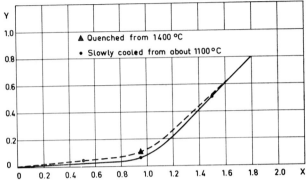

Fig. 3.10. Cation distribution in spinels (Mn_{x-y}, Fe_{1-x+y}) (Fe_{2-y}, Mn_y) O_4 (after Rieck et al. [1966]).

The quantity y, which represents the cation exchange of sites, tends to zero for $x \leq 1$ and increases linearly with $x - 1$ for $x > 1$. Compare Fig. 3.9, which shows qualitatively similar results for some other cations, but the results are not based on experiments of comparable reliability.

From this very condensed review of the various properties of spinel structures, involving transition metals, it seems to be very hard to explain everything in terms of a simple theory, because of the complexity of inter-actions. The most promising compounds are therefore pure ionic spinel structures. For this reason a careful study of the spinels $MgO-Al_2O_2$ has been done in the past and a comprehensive review of these investigations shall be given here.

4.3. Spinels $MgO \cdot xAl_2O_3$ $(1 \leq x \leq \infty)$

The binary system $MgO-Al_2O_3$ contains 3 stable compounds, namely MgO (rock-salt structure), $MgAl_2O_4$ (spinel) and α-Al_2O_3 (corundum structure). MgO can take up a certain excess of Al_2O_3 at high temperatures; the precipitation of $MgAl_2O_4$ at low temperatures has been reported by Stubican et al. [1965]. The spinel structure solves a considerable excess in

References p. 177

Al_2O_3. The limit of solubility is given in Fig. 3.11, which has been taken from the paper by Saalfeld et al. [1957]. As this limit has been determined by heat treatment of samples with precipitated Al_2O_3, the limit of solubility given represents an upper limit. As homogenization might be inhibited, the

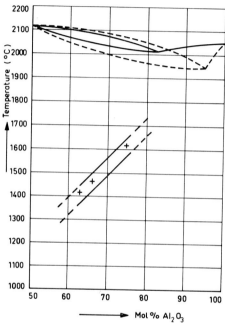

Fig. 3.11. Limit of solubility of Al_2O_3 in $MgAl_2O_4$-crystals (after Saalfeld et al. [1957]).

equilibrium line may lie at lower temperatures than indicated in Fig. 3.11. If a rough extrapolation is correct, the solubility of Al_2O_3 in the spinel lattice should be practically zero at temperatures below 1000°C.

Besides α-Al_2O_3 a couple of other metastable modifications of Al_2O_3 is reported, γ-Al_2O_3 is just one of them. All these structures are hole superstructures of close-packed arrangements. It could not be shown that any simple superstructure of this kind exists for the spinel structure. γ-Al_2O_3 seems to be a metastable or even unstable compound which is probably stabilized by the presence of water or other impurities. This may be checked by the observation that mixed crystals of all compositions, synthesized at temperatures above 2000°C according to Verneuil's method, still contain a small, but varying amount of OH or H_2O. Infra-red absorption diagrams show two absorption maxima (Fig. 3.12) corresponding to two energetically different O–H–O bonds. As the oxygens occupy a single lattice complex, this fact seems to be incompatible with the lattice symmetry if (OH) replaces O, according to the reaction

$$O^{2-} \rightleftarrows (OH)^- + e^- .$$

Although there are different O–O distances, which are possible partners of hydrogen bridges, it should be remembered that the parameter u should

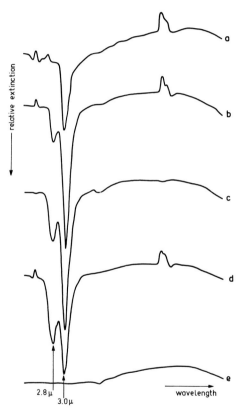

Fig. 3.12. Infra-red absorption of hydrogen-bridges in spinels as a function of Al_2O_3-content. a) spinel MgO: $Al_2O_3 = 1:1.1$; b) spinel MgO: $Al_2O_3 = 1:1.5$; c) spinel MgO: $Al_2O_3 = 1:2.5$; d) spinel MgO: $Al_2O_3 = 1:3.5$; e) corundum (Al_2O_3). It should be noted that the maximum at 2.8 diminishes with decreasing Al_2O_3-contents, but the second maximum at 3.0 does not.

have an important influence on the position of the peaks. In cases, where u is exactly 0.375, the two maxima should coincide. As u decreases from 0.387 to 0.380 for spinels with the composition $MgO:Al_2O_3 = 1:3.5$, a displacement of maxima should be observed. Fig. 3.12 shows that experimentally no displacement is found, as a function of Al_2O_3-contents[1]. Therefore some other models have to be discussed.

Possible molecular reactions are

1) $Al^{3+} + O^{2-} \rightleftarrows Mg^{2+} + (OH)^-$,
2) $Mg^{2+} + 2O^{2-} \rightleftarrows 2(OH)^-$,
3) $Al^{3+} + 3O^{2-} \rightleftarrows 3(OH)^-$

and others.

The first reaction generates no vacancies on cation sites and leads to an increase of MgO contents, which is not observed. The same is true for reaction 3), which diminishes the Al_2O_3 contents. Thus the second reaction or a combination of reactions seem to be most probable. Two hydrogen bonds are formed, one of which may be located in the neighbourhood of the cation-vacancy; the second (OH) is most probably associated with the non-occupied tetrahedral or octrahedral hole. In this way two maxima may be explained. There are still other possibilities of explaining the different OH maxima in the infra-red absorption.

The replacement of cations by protons in ionic crystals seems to be an important factor for nonstoichiometric compounds. Various authors report on many observations of this kind, the details of which cannot be discussed here. But it should be pointed out that corundum crystals, synthesized under similar conditions, often do not show any OH-content, when the method of infra-red absorption is applied (Fig. 3.12). Thus the exchange of cations seems to be one of the important reasons for forming hydrogen bridges. As the observations of hydrogen bridges is rather independent of exsolution and excess in Al_2O_3 (compare Fig. 3.12), molecular reactions, as indicated above, are the most reasonable explanation for the existence of hydrogen bridges. The cation distribution, associated with an excess in Al_2O_3, has been studied by Jagodzinski et al. [1958]; it could be shown that the replacement process can practically be given by the equation

$$3Mg^{2+} \rightleftarrows 3Al^{3+} + \square_0^{3-}$$

[1] I thank Dr. Ollendorf for kindly taking the infra-red absorption spectra of the samples, shown in Fig. 3.12.

where $\square_0$ symbolizes a void on B-sites. Hägg and Söderholm [1935] by density measurements showed that the excess in alumina is realized by voids on the cation sites (cation defect structure). This fact may be understood qualitatively with the aid of Fig. 3.2; the cation chains of the B-sites have a very small distance, much smaller than the minimum A–A distances or A–B distances of the structure. But formally this cannot be understood, using the averaged lattice energies as calculated by Verwey et al. [1948], because of the influence of the parameter μ.

Fig. 3.12 shows clearly that the height of the low frequency absorption maximum decreases with decreasing Al_2O_3-contents, becoming zero for the ideal composition $MgO:Al_2O_3 = 1:1$. This fact indicates that one hydrogen bridge is associated with cation vacancies on B-sites, which are missing for this ideal composition. Therefore it has to be assumed that a couple of replacement processes determines the (OH)-contents in spinels. The second maximum is most probably due to (OH)–O in the octahedral holes, which should have a completely different O–H–O-distance, corresponding to the low frequency maximum.

The replacement reaction for spinels supersaturated in Al_2O_3, leads to an increase of the averaged electrostatic energy, according to Fig. 3.6. The replacement of Mg^{2+} by Al^{3+} decreases the parameter which leads to an unfavourable energy on two reasons:

1) The decrease of u causes an increase of the averaged energy.
2) The increase of averaged charge on A-sites leads to another unfavourable contribution for the values of u in question here (compare Fig. 3.6).
3) The influence of the change of lattice constants is very small only, although a decrease of lattice constants is observed, which is less than 1%.

This unfavourable energy may be compensated by short-range order contribution to the lattice energy.

According to the equations, given in Section 3.2, we can calculate the electrostatic contribution of possible short-range order in the following way: We first consider a small concentration of voids; each void, which is formally a 3^- charge, attracts the replaced 3 Mg^{2+}. If the electrostatic energy of averaged structure is considered to be known, its calculation can be given, according to equations (3.7), (3.8).

Let c be the concentration of voids. The averaged charge on the B-sites is $q_B = 3^+(1-c)$ correspondingly the averaged charge on the A sites is

$$q_A = 2^+(1-3c) + 3c\cdot 3^+ = 2^+(1+\tfrac{3}{2}c) .$$

The energy change caused by these variations of q_A and q_B is unimportant in case of small concentrations. The short-range order contribution is considerably larger and can be calculated with the aid of the following equation

$$U_m = \sum_{v,\mu} \left(\frac{\eta_v \cdot \eta_\mu}{|r'_\mu - r'_v|} - \frac{q_v \cdot q_\mu}{|r_\mu - r_v|} \right). \tag{3.18}$$

In eq. (3.18) η_v and η_μ represent the actual charges of the complete mistake, including the distances r'_v, r'_μ, differing from the distances r_v, r_μ, of the averaged structure. The new distances r'_v, r'_μ may be found by studying the diffuse background with the aid of diffraction methods. The diffuse X-ray intensities are given by a very similar formula (Haefner et al. [1966])

$$I_d = C|\sum_s \exp(2\pi i h r_s^0) \sum_m \exp(2\pi i h \Delta_{st_m}^m)(f_{t_m} - f_s)$$

$$+ \sum_s f_s \exp(2\pi i h r_s^0) \left(\sum_m \exp(2\pi i h \Delta_{ss}^m) - 1 \right)|^2 \tag{3.19}$$

where h = reciprocal vector (no integer),
r_s^0 = corresponds to the ideal position vectors,
$\Delta_{st_m}^m$ = displacement of replaced atom of kind s by atom of kind t in the mth cell (only one t for each m),
Δ_{ss}^m = corresponding displacement of unreplaced atoms,
f_s, f_{t_m} = scattering factor of atoms s, t_m.

The first sum comprehends replaced and displaced atoms and the second one displaced atoms only. The evaluation of eq. (3.19) in terms of a model is not difficult, if a suited computer programme is available; but a rather well established model has to be found, because the variation of parameters involves an amount of calculating as formidable as the solution of a complicated structure by trial and error methods.

By use of equation (3.18) the minimum of electrostatic energy of certain models can be used to find the most probable arrangement of atoms of the defect. Jagodzinski et al. [1967] followed this procedure and got the reasonable result that a certain cation distribution around a vacancy $\Box_0$ was most probable (compare Fig. 3.13). Their results were applicable for the approximate positions of cations; the displacements of the anions cannot be found

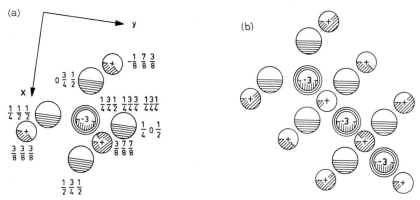

Fig. 3.13. Model of cooperating vacancies $\Box_0$ and replaced Mg^{2+} (by Al^{3+}) of spinels, supersaturated in Al_2O_3. (a) A single group of defects, (b) Several cooperating groups of defects (chains into the [110]-direction).
The vacancy at $\frac{1}{4} \frac{3}{4} \frac{1}{2}$ (a) has six oxygen neighbours; two of them are below and above the vacancy ($\frac{1}{4} \frac{3}{4} \frac{1}{4}$ and $\frac{1}{4} \frac{3}{4} \frac{2}{4}$). Oxygen = large spheres, vacancies = small spheres signed -3.
Mg-sites replaced by Al: small spheres signed $+$, the shaded area of the circles gives the heights of positions.

very easily because the repulsive forces and the polarization of the anions play an important role. Therefore the position of anions could only be estimated.

The general result of this calculation was a displacement of oxygens towards the 3 Mg-positions, replaced by Al, the remaining 3 oxygens should be less displaced, because of the repulsive forces of Mg, but as we are concerned with a statistical spinel, we assumed, as a first approach to the structure, equal averaged displacements for all oxygen neighbours. It is quite obvious that "excited states" of the cooperating group of defects (a void on the B-sites and 3 surrounding replaced A-sites should also be taken into account). This is possible and leads to a small increase in energy. The minimum amount of energy was found for a pyramidal arrangement of the 3 replaced A·sites, but other arrangements lead only to a slight increase in energy. But a considerable amount of energy was found, if one of the replaced A-sites was displaced to the nearest position but one. Therefore in principle a set of possible states has to be discussed for a more accurate calculation of diffuse intensities. With the aid of this method a first approach

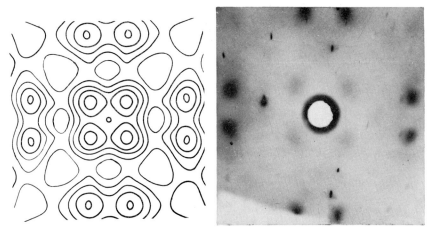

Fig. 3.14 (a). Observed and calculated diffuse intensities according eq. (3.19). Calculation of diffuse scattering of a single group, according Fig. 3.13 (a), compared with a spinel of composition $MgO : Al_2O_3 = 1:1.2$. Laue picture in the [100]-direction. Crystal monochromatized X-rays (Mo-K_α-radiation).

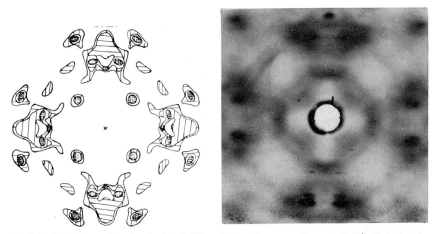

Fig. 3.14 (b). Observed and calculated diffuse intensities according eq. (3.19). Calculated model of two groups, according Fig. 3.13, compared with a spinel 1:3.5. Laue picture into the [100]-direction.

to the interpretation of the diffuse diffraction pattern could be given. Fig. 3.14(a) shows a picture, which is in fairly good agreement with a synthetic spinel with a molar ratio 1:1.2. But a small increase in Al_2O_3 content gives

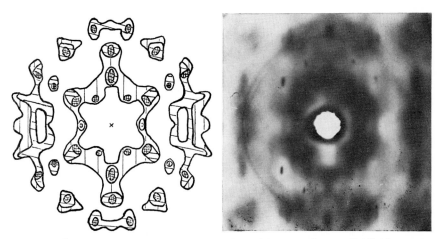

Fig. 3.14 (c). Observed and calculated diffuse intensities according eq. (3.19). The same conditions as indicated for Fig. 3.14 (b), but [110]-direction of the incident X-ray beam.

a much more complicated pattern of the diffuse scattering, which cannot be explained in terms of this simplified model.

The reader may believe that the accurate parameters of the atoms can be found by refinement of the structural model; this is possible in principle, but we didn't apply this method, because of the following arguments:

The intensities of the diffuse scattering are much lower than the intensities of main reflections (on the average 10^{-3} less intense!). In order to carry out a 3-dimensional refinement, a subdivision of the unit cell into at least 5^3 cells of the normal spinel in reciprocal space is necessary. In order to get a sufficiently accurate parameter, measurements up to $h, k, l \approx 10$ are necessary. Bearing in mind that at least one octant in reciprocal space should be measured, in order to have some few equivalent measurements, $10^3 \times 5^3 = 1.25 \times 10^5$ data are necessary. Assuming that the intensity of a main reflection needs roughly a second, 10^3 seconds are needed for the diffuse reflection. Thus 1.25×10^8 seconds are sufficient for the whole experimental procedure. This means roughly that four years are necessary for the experimental work. This time exceeds that required for the intensity measurement of a protein-structure considerably. For this reason it seems doubtful whether a complete structure determination of the short-range order problem is justified if the structure involved is more complex than usual metal structures or some AB-compounds. The interpretation, based on a smaller number of experimental measurements, seems to be acceptable.

In order to determine the short-range order of spinels with a molar ratio $MgO:Al_2O_3$ larger than $1:1.2$, we first tried to check, whether the interaction energies between two adjacent faults is sufficient to explain the diffuse pattern in terms of a cooperative phenomenon.

A similar calculation, using eq. (3.18), gave a favourable energy for two cooperating groups of defects in the [110]-direction, according to Fig. 3.13(b). The spatial model apparently gives an increase of the distance of two cooperating voids in the [110]-direction and a decrease of distances to the surrounding Al^{3+} replacing Mg^{2+}, embedding the voids, including their displaced oxygen neighbours. This fact causes considerable strains of the surrounding matrix crystal, involving an unfavourable strain energy considerably larger than the strain energy associated with a single group of defects. These few conclusions in terms of a very simple model, and some additional calculations of the interaction energies of 3 groups of defects lead to the conclusion that the attachment of a third group is not very probable. All these considerations have a qualitative basis only; the next step for a better agreement should be the calculation of the distribution function of the group of defects; this may be done by introducing sequence probabilities p_n, giving the probability that n attached groups will be followed by another one.

Calculations showed that

a) a favourable energy in the [110]-direction is very small, p_3 may become nearly zero, because of the unfavourable strain energy, mentioned above;

b) a distribution function, primarily containing pairs, is possible, if a suited set of α_n is chosen. The probability of finding a complex of n groups is

$$p_n = \alpha_0 \alpha_1 \ldots (1 - \alpha_n) \qquad (3.20)$$

where $\alpha_0 =$ equal to the concentration of defects. Using eq. (3.20) and a suited set of α_i, the problem of X-ray diffraction may again be calculated by an incoherent summation of the various groups because random distribution of groups may be assumed:

$$I_d = \sum_n \alpha_n I_d(n) \qquad (3.21)$$

where $I_d(n)$ has to be calculated according to eq. (3.19) with the model of the nth grouping.

The application of this method is not restricted to the one-dimensional case because the groups may have any three-dimensional arrangement. The main difficulty is finding the theoretical background for the suited set of α_n. The best agreement has been obtained with a relative mean displacement $\bar{\Delta}$ of oxygens from the positions of the averaged structure of $\bar{\Delta} = 0.075$ Å while, for a single group the smaller value of 0.015 gave the best agreement. The difference may well be understood in terms of the discussion, concerning increasing strains with increasing number of cooperating groups.

The parameter $\bar{\Delta}$ has first been determined by assuming that all displacements correspond to the change in the parameter u of the cubic space group, which may be estimated from the paper by Jagodzinski et al. [1958]. This is safely incorrect, because of the lower symmetry of the cooperating group. Generally the displacement of the oxygens should be subject to symmetry-conditions, given by the grouping. This symmetry is only $2/m$ with the twofold axis parallel to [110] of the spinel lattice. Thus the oxygens should be displaced without any restriction. Only those atoms lying in the symmetry plane should be fixed in that plane.

Therefore it may be concluded that the model discussed above can only give an approximate agreement with the observed diffuse intensities; the refinement of intensities with free oxygen parameters should result in a much better agreement between calculation and observed intensities. But as the calculated diffuse intensities for diffraction patterns, taken in the [110]-direction of the incident beams is nearly as good, we did not try to carry out a refinement. The distribution of disordered atoms or groups of atoms is without any doubt one of most important factors giving the physical and chemical properties of the nonstoichiometric compound and it could be shown that energy calculation – even in a semi-quantitative way – is a valuable aid in the solution of problems, which cannot be found by the use of diffraction data alone.

What happens in case of very high concentration of defects in the mixed crystal? To answer this question, we should know the limit of solubility of Al_2O_3 – in the spinel lattice. The investigation by Saalfeld and Jagodzinski [1957] (compare Fig. 3.11) shows that at temperatures about 1600°C a spinel with a molar ratio 1:4 is in equilibrium. The concentration of voids on B-sites has increased up to 12%. As the short-range order, discussed

References p. 177

above, cannot lead to a three-dimensional superstructure, other models of cooperating defects have to be suggested. They should have an unfavourable energy for small groups but be able to form a completely ordered three-dimensional structure. The formation of a critical nucleus, which is able to continue growth, is connected with an activation energy.

This has been confirmed by Saalfeld et al. [1957], who found a complicated superstructure of the original spinel lattice, as a precipitate of Al_2O_3 or a compound very near to this composition at higher temperatures. The cation distribution is completely different from that of the spinel structure, supersaturated in Al_2O_3.

The structure has been solved by Jagodzinski [1960], without the application of a refinement procedure, and it may well be that some corrections are necessary. This intermediate structure is obtained by annealing the spinel between 900–1100°C. This kind of precipitation behaviour is very interesting, because the suggestion by Rinne [1928], that this intermediate state of precipitation with the complicated superstructure is γ-Al_2O_3 in a spinel matrix could not be confirmed. It is this very reason why γ-Al_2O_3 is most probably an unstable modification of Al_2O_3. On account of the very close relationship of the spinels and the γ-structure of Al_2O_3, precipitation of γ-Al_2O_3 in supersaturated spinels, should be normally expected as an intermediate metastable structure. In spite of a very careful study at various temperatures, no indication of γ-Al_2O_3 as precipitation has been found so far, although the lattice constant of γ-Al_2O_3 and $MgAl_2O_4$ differ less than 15%. The intermediate monoclinic phase, described above, has a superstructure of cations, preferably on octahedral sites, therefore this structure may be interpreted in terms of the rock salt-arrangement with a superstructure of holes on cation sites.

As has been pointed out above, spinels, grown from a melt with an excess in Al_2O_3 compared with the ideal composition, grow without difficulty while the 1:1 composition, which corresponds to the ordered $MgAl_2O_4$ does not form well-grown crystals. This seems to be curious, but may be understood by consideration of the entropy. It has been shown that at 900°C the cation distribution no longer corresponds to the "normal" spinel, therefore the maximum entropy for the spinel of the composition AB_2O_4 at very high temperatures (above 2000°C) is

$$S = k \ln \frac{(3n)!}{n!\,(2n)!}.$$

(3.22)

For the spinel, containing a surplus of Al_2O_3 the maximum entropy at high temperatures is

$$S = k \ln \frac{(3n)!}{[n(1-x)]! \, [\frac{1}{3}nx]! \, [2n(1+\frac{1}{3}x)]!} \, . \qquad (3.23)$$

The maximum configuration entropy, according to eq. (3.23) is given by

$$\frac{dS}{dx} = 0 \, , \qquad (3.24)$$

and may be evaluated with the aid of Stirling's formula.

This leads to maximum entropy for $x \approx \frac{1}{3}$, which corresponds to a molar ratio $MgO:Al_2O_3 = 1:1.5$, which is considerably lower than the concentration range where most stable spinels are found. There should be an additional factor for the explanation of this fact. It is not impossible that the free octahedral holes (02), which are larger than occupied ones (01), may be filled by a possible crystal chemical reaction

$$Al_T \rightarrow Al_{02} \text{ or } Al_{01} \rightarrow Al_{02}$$

leaving voids on the tedrahedral (or octahedral) positions; that means the structure has a tendency towards a statistical rock salt-structure with $\frac{2}{3}$ filled octahedral positions. But this could not be confirmed by X-ray powder diffraction data up to 1600°C.

As this possible transformation is not characterized by a critical temperature, this observations cannot exclude the possibility of the cation distribution mentioned above. A change of the equilibrium composition to spinels rich in alumina will occur if there is a site preference for Al. This may be true as long as the spinel structure does not degenerate into a statistical spinel with all cations on octahedral sites, but this has not yet been observed. It is not possible to give a detailed explanation of this composition – dependent on a change of equilibrium conditions, as long as the parameter u and cation distribution has not been determined, but it should be emphasized that the change in composition of definite compound by entropy is an interesting feature of nonstoichiometry.

Another possible transition has to be discussed, although it seems unlikely that it is important in the case of Mg-Al-spinels. There is a second structure possible by occupying the equivalent lattice complexes translated by $\frac{1}{2}\frac{1}{2}\frac{1}{2}$ from the positions discussed in this paper. As both structures are symmetrically equivalent, the existence of domain structures at low temperatures is possible and a transition with a critical temperature at very high

References p. 177

temperature cannot be excluded. In case of the correct composition AB_2C_4 the site probabilities $p_{i|v}$ are then equal for all cation sites. But there is no indication so far that this possible transformation plays an important role although domain structures cannot be definitely excluded.

5. Conclusions

This paper does not aim to give a complete report on the crystallography of nonstoichiometric compounds. The literature cited gives a selection of papers, important for certain types of spinels, but cannot be considered representative of the complete field of stoichiometry, which covers a huge number of publications. Even the literature on spinels, which comprehends several hundreds of papers, could not be cited completely here. For this reason it seemed desirable to give a very detailed review of a small field established by a maximum number of reliable experiments. It could be shown that the solution of the diffraction problem of disordered crystals is closely related to nonstoichiometry. The static structure problem can be solved with moderate difficulty in the case of simple compounds; complicated structures involve increasing difficulties and often other methods have to be applied in order to get an unique solution of the statistical structure, for which short-range interactions are of general importance.

It is obvious that the dynamic behaviour of crystals is influenced by the disorder phenomena, correlated with nonstoichiometry. Theoretical calculations and some experimental measurements indicate new vibrational modes and a change of frequencies of existing ones. As the experimental results and their theoretical interpretation is poor, there has been no attempt to review this part of nonstoichiometry. The influence of nonstoichiometry on physical properties is discussed in this book elsewhere, but it should be pointed out here that the correlation between structural research and physical properties should be based on more reliable experimental results, which cannot be given by the application of the powder technique in most cases. This is the reason why the investigation of single crystals resulted in a full understanding of the physical and chemical behaviour of the Mg-Al spinels. Further experiments are necessary for a better understanding of other groups of nonstoichiometric crystals.

References

ALPERIN, H. A., 1962, J. Phys. Soc. Japan. **17**, B III, 57.

BLASSE, G. 1964, Crystal Chemistry and some magnetic Properties of mixed Metal Oxydes with Spinel Structure, Thesis Leiden.

BRAGG, W. L. and E. J. WILLIAMS, 1934, Proc. Roy. Soc. A **115**, 699.

BRAVAIS, A. 1850, J. de l'école polytechnique, Paris **19**, 1.

BRUN, E. and S. HAFNER, 1962, Z. Krist. **117**, 37.

DE BOER, F., J. H. VAN SANTEN and E. J. W. VERWEY, 1950, J. Chem. Phys. **18**, 1032.

DUNITZ, J. D. and L. E. ORGEL, 1957, Phys. Chem. Solids **3**, 318.

FRANKENHEIM, M. L., 1856, Poggendorffs Annalen, XCVII.

GOERTER, E. W., 1954, Philips Res. Rept. **9**, 295.

HAEFNER, K. and H. JAGODZINSKI, 1966, Acta Cryst. **20**, 17.

HAFNER, S., 1960, Schweiz. Mineral. Petrog. Mitt. **40**, 207.

HAFNER, S. and F. LAVES, 1961, Z. Krist. **115**, 321.

HASTINGS, J. M. and L. M. CORLISS, 1956, Phys. Rev. **104**, 328.

HUANG, K., 1947, Proc. Roy. Soc. A **190**, 102.

JAGODZINSKI, H. and H. SAALFELD, 1958, Z. Krist. **110**, 197.

JAGODZINSKI, H., 1960, Z. Krist. **119**, 388.

JAGODZINSKI, H., 1963, Crystallography and Crystalperfection (Academic Press, London and New York) p. 177 ff.

JAGODZINSKI, H. and K. HAEFNER, 1967, Z. Krist. **125**, 188.

KLEINSTÜCK, K., E. WIESER, P. KLEINERT and R. PERTHEL, 1965, Phys. Stat. Solids **8**, 271.

KURNAKOW, S. N., 1914, Z. Anorg. Allgem. Chemie **88**, 109.

LENNÉ, H. U., 1961, Z. Krist. **115**, 297.

MANDELKORN, L., 1964, Non-stoichiometric Compounds (Academic Press, New York and London).

McGILLARY, G. H. and B. STRIJK, 1946, Physica **11**, 369.

McCLURE, P. S., 1957, Phys. Chem. Solids **3**, 311.

RIECK, G. D. and F. C. M. DIESSENS, 1966, Acta Cryst. **20**, 521.

RINNE, F., 1928, N. Jahrb. Mineralogie (A) **58**, 43.

ROBINSON, K. and G. W. BRINDLEY, 1949, Proc. Leeds Phil. Soc. **5**, 109.

SAALFELD, H. and JAGODZINSKI, 1957, Z. Krist. **109**, 87.

SMIT, J., F. K. LOTGERING and VAN STAPELE, 1962, J. Phys. Soc. Japan **17**, B-1, 268.

STUBICAN, V. S. and R. ROY, 1965, Phys. Chem. Solids **26**, 1293.

VAN SANTEN, J. H., 1950, Philips Res. Rept. **5**, 282.

VERWEY, E. J. W., F. DE BOER and J. H. VAN SANTEN, 1948, J. Chem. Phys. **16**, 1091.

WAGNER, C. and W. SCHOTTKY, 1930, Z. Physik. Chem. (Leipzig) **B111**.

WEISZ, A., 1958, Chem. Ber. **91**, 1487.

YAMZIN, J. J., N. V. BELOV and Y. NOZIK, 1962, J. Phys. Soc. Japan **17**, B 111, 55.

CHAPTER 4

METASTABLE NONSTOICHIOMETRIC COMPOUNDS WITH ORDERED VACANCIES

J. C. JOUBERT, G. BERTHET AND E. F. BERTAUT

Laboratoire d'électrostatique et de physique du metal, Grenoble, France

1. Introduction

This paper describes an original method for the synthesis of various new nonstoichiometric compounds with ordered vacancies. According to this method, a lot of lithium containing oxides can be used as starting materials to prepare nonstoichiometric lithium free compounds.

The general chemical reaction can be sketched as below:

$$2Li^+ \rightarrow M^{2+} + \square$$

where M^{2+} is a bivalent cation and the sign $\square$ stands for a vacancy.

In this exchange reaction process, the basic structure of the starting oxides remains unchanged; however in most cases an order appears between the vacancies and the exchanged bivalent cations on the lithium sites.

While the new vacancies containing compounds are quite stable at room temperature, they generally collapse when the temperature is raised at a few hundred degrees C.

2. Principle of the solid state exchange reaction

2.1. Examples of exchange reactions in mineral chemistry

It is known for a long time that zeolite materials possess the important property of exchanging their alkaline cations with those of a basic surrounding chemical agent, according to the reaction

$$Ca \rightleftarrows 2Na \text{ (or } 2K, 2NH_4 \text{ etc...)}.$$

References p. 208

The Feldspars can also undergo such exchange reactions, as is the case in the next instance (Wyart et al. [1956a, b])

$$KAlSi_3O_8 + NaCl \xrightleftharpoons{t > 350\,°C} NaAlSi_3O_8 + KCl .$$
orthoclase   albite

2.2. Development of the method

In zeolites as well as in feldspars, we noted that the frame of the starting materials was unaltered during the exchange process. We tried to extend the exchange method to other kinds of oxides.

In order to prepare new nonstoichiometric compounds, we imagined to use as starting materials lithium containing oxides and to realize the exchange:

$$2Li^+ \rightarrow \square + M^2 . \tag{4.1}$$

As our aim was to preserve the starting material's frame, we first fixed our choice on spinel oxides, in which the oxygen close packing secured a very steady substructure.

After a few negative tests, we succeeded in realizing the following exchange reaction:

$$(Ge_3/Li)LiZnO_8 + ZnSO_4 \xrightarrow{500\,°C} (Ge_3/\square)Zn_2O_8 + Li_2SO_4 . \tag{4.2}$$

The starting material was a spinel with $4 \times (3Ge^{4+} + 1Li^+)$ cations on the octahedral sites and $4 \times (1Li^+ + 1Zn^{++})$ cations on the tetrahedral sites. On the octahedral sites, an $(1:3)$ ordered distribution between Ge^{4+} and Li^+ was easily detected by X-ray analysis. The same ordered distribution between Ge^{4+} and the vacancies was observed on the octahedral sites of the exchanged compound.

The solid state exchange reaction was performed in keeping a mixture of $Ge_3Li_2ZnO_8$ and $ZnSO_4$ (in excess) for several days at about 500°C. The vacancies containing compound was easily separated from zinc (in excess) and lithium (just formed) sulfates by abundantly washing the final mixture in pure cold water. X-ray analysis and intensity calculations showed the compound to be pure $(Ge_3/\square)Zn_2O_8$.

The new defect spinel is quite stable at room temperature. However, when heated for a few minutes above 600°C, it decomposes according to the reaction (4.3):

$$(Ge_3/\square)Zn_2O_8 \xrightarrow{t=600\,^\circ C} GeZn_2O_4 + 2GeO_2 .$$ (4.3)

defect spinel   phenakite

By annealing the (4.3) reaction product for several days at 500°C, we did not observe any change in the intensities on the X-ray pattern: no trace of the defect spinel could be detected. The reaction seems thus, to be irreversible. Moreover, by maintaining the defect spinel $(Ge_3 \square)Zn_2O_8$ at 550°C for a long time – 3 months for instance – we found that about half of the starting material was decomposed according to the reaction (4.3). It is likely that keeping the compound at even lower temperatures, one would get the same result, but after a much longer period. At room temperature, this period becomes infinitely long and the vacancies containing spinel appears to be quite stable.

As it can be seen on the DTA curve corresponding to reaction (4.3) a very exothermic signal is detected at about 600°C. We will call it "temperature of rapid decomposition", but it must be pointed out that the compound is metastable at any temperature.

Other exchange reactions have been attempted with success (Joubert [1965]). We shall describe a few examples in the next chapter, but we want to remind first of all some requirements which have to be fulfilled in order to obtain a positive result:

The starting material and the defect exchanged compound have to be very insoluble in water and unhydrolysable. Thus they can easily withstand the prolonged washing which is necessary to eliminate any trace of exchange agent.

The exchange is performed in the good way only if the free energy of formation of the bivalent salt is lower – at the reaction temperature – than the free energy of formation of the lithium salt, as in reaction (4.2):

$$\left.\begin{array}{l} \Delta F = Zn\,SO_4 = 230\ kcal/mole \\[2mm] \Delta F = Li_2SO_4 = 337\ kcal/mole \end{array}\right\} \text{at } 400\,^\circ C.$$

The defect compound is generally unstable if the ionic radii of the exchanged cations (Li^+ and M^{2+}) are very different. As a matter of fact, the largest bivalent cation we could substitute to Li^+ (0.68 Å) is Cd^{2+} (0.97 Å). With larger cations, the exchange process does take place but the vacancies containing compound collapses as soon as it is formed (the starting compound's frame is not preserved).

Sulfates are very convenient exchange agents: their average decomposition temperature is about 700°C, compared to 600°C for chlorides and 200°C for nitrates.

Moreover, they are very soluble in cold water and thus easily eliminated by washing.

3. Examples of metastable vacancies containing compounds prepared by solid state exchange reaction

3.1. Experimental method

A few grams of a well crystallized sample of the starting material is mixed with a large amount of the exchange agent (in the ratio of one part to about ten). The mixture is squeezed at a pressure of a few thousands psi into a pellet to increase the contact between the grains. Then the pellet is placed in a furnace and slowly heated up to an experimentally determined temperature, which must be high enough to provide a rapid exchange without destroying the defect compound. After keeping the pellet for one or two weeks at that temperature, the resulting mixture is abundantly washed with pure water. The exchanged compound is then allowed to decant in the container.

By X-ray analysis, we can test the progress of the exchange reaction. If the exchange is not complete – that is to say the exchanged compound still contains some lithium – everything has to be done over again from the very beginning, until no significant progress can be observed any more. In most of the cases, a one week run, is long enough to get a total exchange, but sometimes it is necessary to carry out two successive runs.

3.2. New metastable defect compounds with ordered vacancies

After the encouraging result obtained for the synthesis of the compound $(Ge_3 \square)Zn_2O_8$, we extended the method with success to prepare other isostructural metastable defect spinels. The general formula of the starting lithium containing spinels was $(T_3^{4+}/Li^+)Li^+M^{2+}O_8$ where T^{4+} stands for the tetravalent cations Ge^{4+}, Ti^{4+}, Mn^{4+} and M^{2+} for one of the bivalent cations Zn^{2+}, Co^{2+}, Mg^{2+}, Mn^{2+}, Ni^{2+}, Cd^{2+}. In each of these spinels we were able to detect an ordered distribution of the lithium and the tetravalent cations on the octahedral sites (Durif et al. [1962], Joubert et al. [1964]). The general solid state exchange reaction can be schematized as below:

$$(T_3^{4+}/Li^+)Li^+M^{2+}O_8 + \underset{\text{in excess}}{M'^{2+}SO_4}$$

$$\xrightarrow[\text{high temperature}]{\text{in solid state}} (T_3^{4+}/\square)M^{2+}M'^{2+}O_8 + Li_2SO_4 \ .$$

Table 4.1 summarizes the experimental conditions that allowed us to prepare the new metastable compounds:

Table 4.1

Preparation of metastable spinels by solid state exchange reaction

Starting materials	Exchange agent	Exchange temperature	Duration	Lacunary compounds	Temperature of rapid decomposition
$(Ti_3Li)(CoLi)O_8$	$CoSO_4 \cdot 7H_2O$	480°C	a week	$(Ti_3 \square)(Co_2)O_8$	570°C
$(Ti_3Li)(MnLi)O_8$	$MnSO_4 \cdot H_2O$	310°C	a week	$(Ti_3 \square)(Mn_2)O_8$	330°C
$(Ti_3Li)(ZnLi)O_8$	$ZnSO_4 \cdot 7H_2O$	500°C	a week	$(Ti_3 \square)(Zn_2)O_8$	900°C
$(Ti_3Li)(ZnLi)O_8$	$CdSO_4 \cdot \frac{8}{3}H_2O$	450°C	a week	$(Ti_3 \square)(ZnCd)O_8$	550°C
$(Ti_3Li)(ZnLi)O_8$	$MnSO_4 \cdot H_2O$	480°C	a week	$(Ti_3 \square)(ZnMn)O_8$	550°C
$(Ge_3Li)(ZnLi)O_8$	$ZnSO_4 \cdot 7H_2O$	500°C	a week	$(Ge_3 \square)(Zn_2)O_8$	600°C
$(Ge_3Li)(CoLi)O_8$	$CoSO_4 \cdot 7H_2O$	500°C	a week	$(Ge_3 \square)(Co_2)O_8$	700°C
$(Ge_3Li)(ZnLi)O_8$	$MnSO_4 \cdot H_2O$	580°C	a week	$(Ge_3 \square)(ZnMn)O_8$	630°C
$(Ge_3Li)(ZnLi)O_8$	$MgSO_4 \cdot 7H_2O$	680°C	a week	$(Ge_3 \square)(ZnMg)O_8$	720°C
$(Mn_3Li)(LiZn)O_8$	$ZnSO_4 \cdot 7H_2O$	300°C	a week	$(Mn_3 \square)(Zn_2)O_8$	340°C

In each case, the X-ray analysis and the intensity calculations show the exchange to be complete. This is easy to check because of the considerable variation in the intensities of some characteristic lines (111, 220, 400, etc...) as shown by the Tables 4.2, 4.3, 4.4, 4.5, 4.6.

References p. 208

Table 4.2

Comparison between the X-ray patterns of $(Ge_3Li)LiZnO_8$ and $(Ge_3 \square)Zn_2O_8$ ($\lambda K\alpha Cu$)

hkl	$Ge_3Li_2ZnO_8$		$(Ge_3 \square)Zn_2O_8$	
	$I_{obs.}$	$d_{obs.}$	$I_{obs.}$	$d_{obs.}$
110	23	5.79	25	5.81
111	45	4.73	7	4.75
210	34	3.66	39	3.68
211	17	3.35	21	3.36
220	20	2.89	57	2.91
221	1	2.73	0.5	2.74
310	4	2.59	4	2.61
311	100	2.47	100	2.49
222	11	2.36	8	2.38
320	5	2.27	3	2.28
321	5	2.19	3	2.20
400	44	2.05	20	2.06
322–410	0.5	1.984	1	1.989
411–330	4	1.931	4	1.938
331	3	1.877	0	
420	0			
421	8	1.786	7	1.791
332	0		0	
422	7	1.672	17	1.678
430	0		0	
510–431	6	1.606	6	1.610
511–333	42	1.577	59	1.580
520–432	6	1.522	7	1.526
521	3	1.497	4	1.500
440	58	1.448	54	1.450
433–530	2	1.406	2	1.406
531	7	1.384	1	1.384
442	0		0	
610	1	1.347	1	1.347
611–532	3	1.329	3	1.329
620	3	1.296	7	1.296
533	13	1.250	16	1.250

a) Crystallographic analysis

Table 4.7 gives the elementary cubic cells' parameters for the starting spinels materials and the defect exchanged compounds.

Table 4.3

Comparison between the X-ray patterns of $(Ge_3Li)LiCoO_8$ and $(Ge_3 \square)Co_2O_8$ ($\lambda K\alpha Co$)

hkl	$Ge_3Li_2CoO_8$		$(Ge_3 \square)Co_2O_8$	
	$d_{obs.}$	$I_{obs.}$	$d_{obs.}$	$I_{obs.}$
110	5.78	25	5.81	18
111	4.72	86	4.73	16
210	3.65	29	3.67	34
211	3.34	12	3.35	16
220	2.89	10	2.90	29
221	2.73	< 1	2.73	1
310	2.59	2	2.60	3
311	2.47	100	2.48	100
222	2.36	13	2.37	10
320	2.27	5	2.28	3
321	2.19	4	2.19	3
400	2.05	57	2.05	24
410–322	1.935	1	1.993	< 1
411–330	1.929	3	1.937	4
331	1.878	8		0
420		0		0
421	1.786	6	1.792	8
332		0		0
422	1.672	2	1.676	10
430		0		0
510–431	1.606	6	1.611	7
511–333	1.576	40	1.581	39
520–432	1.521	5	1.526	6
521	1.495	3	1.500	3
440	1.448	62	1.453	46

These parameters, together with the X-rays patterns, indicate a close analogy between the two structures.

All these compounds belong to space group $P4_332$ (or $P4_132$) with the following distribution:

$4 \square$ (or $4Li^+$)   in   4(b)

$12 T^{4+}$ in 12(d) with $x \approx \frac{3}{8}$

$8 M^{2+}$ (or $4M^{2+}+4Li^+$) in   8(c) $x \approx 0$

$24 O^{--}$ in 24(e) $x \approx \frac{1}{8}, y \approx \frac{1}{8}, z \approx \frac{3}{8}$

$8 O^{--}$ in   8(c) $x \approx \frac{3}{8}$

References p. 208

Table 4.4

Comparison between the X-ray patterns of $(Ti_3Li)LiCoO_8$ and $(Ti_3 \square)Co_2O_8$ $(\lambda K\alpha Co)$

hkl	$Ti_3Li_2CoO_8$		$(Ti_3 \square)Co_2O_8$	
	$d_{obs.}$	$I_{obs.}$	$d_{obs.}$	$I_{obs.}$
110	5.94	10	5.96	23
111	4.84	23	4.85	4
210	3.75	12	3.76	19
211	3.42	7	3.43	15
220	2.96	13	2.97	50
221	2.80	1	2.80	3
310	2.65	4	2.66	5
311	2.53	100	2.53	100
222	2.41	2	2.42	1
320	2.33	2	2.33	3
321	2.24	1	2.24	2
400	2.09	49	2.10	16
410–322	2.03	1	2.03	1
411–330	1.977	1	1.977	1
331	1.927	1	1.927	1
420		0		0
421	1.829	5	1.829	6
332		0		0
422	1.709	7	1.709	18
430		0	1.674	1
510–431	1.642	1	1.642	1
511–333	1.612	40	1.612	42
520–432	1.555	4	1.555	4
521	1.529	4	1.529	5
440	1.481	56	1.481	51

where T^{4+} stands for Ti^{4+}, Ge^{4+}, Mn^{4+} and M^{2+} for Zn^{2+}, Cd^{2+}, Mn^{2+}, Co^{2+}, Mg^{2+}

b) Thermal stability of the defect spinels

One has to bear in mind that the decomposition temperatures given in Table 4.1 are those of rapid decomposition and that any one of the defect compounds will at last collapse if heated for a long enough period at say 50°C below the listed "rapid decomposition" temperature. A look on Table 4.1 shows how critical is sometimes the choice of the temperature which is necessary to secure good reaction kinetics.

Tanle 4.5

Comparison between the X-ray patterns of $(Ti_3Li)LiMgO_8$ and $(Ti_3 \square)Mg_2O_8$ ($\lambda K\alpha Cu$)

hkl	$Ti_3Li_2MgO_8$	$(Ti_3 \square)Mg_2O_8$
	$I_{obs.}$	$I_{obs.}$
110	25	38
111	100	48
210	35	44
211	23	33
220	10	20
221	1	2
310	6	5
311	81	100
222	1	1
320	3	2
321	2	2
400	85	63
410–322	0	1
411–330	3	2
331	4	1
420	0	0
421	8	9
332	0	0
422	2	6
430	0	0
510–431	4	3
511–333	40	42
520–432	11	7
521	10	9
440	62	64

3.3. New metastable allotropic phases with ordered vacancies

In Section 3.2, we applied the solid state exchange reaction method to the preparation of new defect compounds.

In this section, we shall describe new metastable allotropic phases of already known compounds, obtained by the method.

A second allotropic form of cobalt metatitanate $TiCoO_3$

The usual form of cobalt metatitanate belongs to the well known il-

Table 4.6

Comparison between the X-ray patterns of $(Mn_3Li)LiZnO_8$ and $(Mn_3 \square)Zn_2O_8$ $(\lambda K\alpha Cu)$

hkl	$Mn_3Li_2ZnO_8$		$(Mn_3 \square)Zn_2O_8$	
	$d_{obs.}$	$I_{obs.}$	$d_{obs.}$	$I_{obs.}$
110	5.75	13	5.78	17
111	4.74	33	4.76	7
210	3.65	24	3.67	24
211	3.33	9	3.35	8
220	2.90	39	2.92	70
221	2.73	0		0
310	2.59	1	2.60	1
311	2.46	100	2.48	100
222	2.36	5	2.37	1
320	2.27	2	2.29	2
321	2.19	2	2.20	2
400	2.04	30	2.06	9
410–322	1.986	1		0
411–330	1.926	2		0
331	1.876	2.5	1.883	6
420	1.829	6	1.836	0
421	1.784	3	1.790	4
332		0		0
422	1.671	13	1.678	17
430		0		0
510–431	1.606	2	1.701	4
511–333	1.574	49	1.579	40
520–432	1.520	2.5	1.526	6
521		0		0
440	1.448	57	1.456	43

menite structural type, space group $R\bar{3}$. The hexagonal triple cell contains 6 formula units. The cell parameters are:

$$a = 5.068 \text{ Å}, \quad c = 13.922 \text{ Å}, \quad z = 6.$$

The structure is derived from the corindon type $(\alpha\text{-}Fe_2O_3)$ by the establishment of an order between Ti^{4+} and Co^{2+} cations on the Fe^{3+} sites.

The oxygene frame forms an almost hexagonal close packing.

Starting with the lithium spinel $Ti_4Li_2Co_3O_{12}$ $(a = 8.398 \text{ Å})$ we performed the exchange reaction (Berthet [1968]):

Table 4.7

Cell parameters of the starting spinels and the defect exchanged compounds

Formula	Cell parameter (Å)	$d_x(g/cm^3)$
$(Mn_3Li)LiZnO_8$	8.193 ± 0.002	4.5
$(Ge_3Li)LiZnO_8$	8.190 ± 0.002	5.11
$(Ge_3Li)LiCoO_8$	8.204 ± 0.002	5.06
$(Ti_3Li)LiMgO_8$	8.375 ± 0.005	
$(Ti_3Li)LiZnO_8$	8.372 ± 0.005	3.97
$(Ti_3Li)LiCdO_8$	8.535 ± 0.005	4.25
$(Ti_3Li)LiCoO_8$	8.377 ± 0.005	3.89
$(Ti_3Li)LiMnO_8$	8.440 ± 0.005	3.70
$(Ti_3 \square)Mg_2O_8$	8.390 ± 0.005	
$(Ti_3 \square)Co_2O_8$	8.404 ± 0.005	4.25
$(Ti_3 \square)Mn_2O_8$	8.48 ± 0.01	4.05
$(Ti_3 \square)Zn_2O_8$	8.395 ± 0.005	4.52
$(Ti_3 \square)ZnCdO_8$	8.394 ± 0.005	4.94
$(Ti_3 \square)ZnMnO_8$	8.449 ± 0.005	4.21
$(Ge_3 \square)Zn_2O_8$	8.213 ± 0.002	5.72
$(Ge_3 \square)Co_2O_8$	8.237 ± 0.005	5.51
$(Ge_3 \square)ZnMnO_8$	8.217 ± 0.005	5.58
$(Ge_3 \square)ZnMgO_8$	8.214 ± 0.005	5.35
$(Mn_3 \square)Zn_2O_8$	8.220 ± 0.005	5.06

$$Ti_4Li_2Co_3O_{12} + CoSO_4 \xrightarrow[\text{2 weeks}]{400\,°C} (Ti_4Co\square)Co_3O_{12} + Li_2SO_4 . \qquad (4.4)$$

disordered spinel   ordered spinel

The cell parameter of the cubic allotropic form of $TiCoO_3$ is $a = 8.429$ Å.

The space group is $P4_332$, different from the space group of the starting spinel (Fd3m). This is due to an ordered distribution of vacancies on the octahedral sites.

The following distribution of cations in the elementary cell was determined by X-ray analysis: cubic $TiCoO_3$: Space group $P4_332$

$8\ Co^{2+}$ in 8(c) tetrahedral sites

$\frac{32}{3}\ Ti^{4+} + \frac{4}{3}Co^{2+}$ in 12(d)

$\frac{4}{3}\ Co^{2+} + \frac{8}{3}\square$ in 4(a)

octahedral sites .

This allotropic form is metastable. By heating it a few minutes above 650°C, we get the well known ilmenite form, according to the reaction:

References p. 208

Table 4.8

Comparison between the cristallographic characteristics of the two allotropic phases of $TiCoO_3$ and Fe_2O_3

Formula		Oxygen packing	Cations environment	Space group	Rapid decomposition temperature
Metastable forms	$(Ti_4Co\,\square)Co_3O_{12}$	c.c.p.	octahedral and	$P4_332$ or	650°C
	$(Fe_5\,\square)Fe_3O_{12}$	c.c.p.	tetrahedral	$P4_132$	250°C
Stable forms	$TiCoO_3\alpha$	h.c.p.	octahedral only	$R\bar{3}$	
	$Fe_2O_3\alpha$	h.c.p.		$R\bar{3}c$	

$$\underset{\text{spinel type}}{(Ti_8Co/Co\square\,\square)Co_6O_{12}} \xrightarrow{650°C} \underset{\text{ilmenite type}}{4TiCoO_3}\,.$$

The transformation is very exothermic, as can be seen from the DTA curve, and is completely irreversible. Meanwhile, we note an increase of the density of about 8%: Thus in that case again, the new defect form seems to be metastable.

Let us point out the remarkable analogy between the two allotropic forms of $TiCoO_3$ and the α and γ forms of Fe_2O_3 (see Table 4.8).

A new allotropic form of cobalt orthovanadate

Cobalt orthovanadate $Co_3^{2+}V_2^{5+}O_8$ belongs to the space group Abam; the cell parameters are:

$$a=8.310\ \text{Å}, \quad b=6.03\ \text{Å} \quad c=11.48\ \text{Å}.$$

The oxygen network forms a cubic close packing in which Co^{2+} cations occupy octahedral sites and V^{5+} cations tetrahedral sites.

Starting with the lithium containing spinel $(Co^{2+}Li^+)V^{5+}O_4$ $(a=8.276$ Å) we succeeded in preparing by solid state exchange reaction a new metastable form of cobalt orthovanadate with cubic symmetry (Joubert [1965]).

Table 4.9

Diffractograms of VCoLiO$_4$ and V$_2$(Co$_3$ □)O$_8$ (λKαCo)

hkl	VCoLiO$_4$ Space group Fd3m		(Co$_3$ □)V$_2$O$_8$ Space group P4$_3$32	
	d_o	I_o	d_o	I_o
110			5.88	17
111	4.75	4		
210			3.71	19
211			3.39	18
220	2.92	60	2.94	27
221			2.77	4
310			2.63	6.5
311	2.49	100	2.51	100
222		0	2.40	2
320			2.31	5
312			2.22	2
400	2.063	23	2.076	32
410				0
322				
411				0
330				
331		0		0
420				0
421			1.814	9
332				0
422	1.685	23	1.697	13
430				0
510				0
431				
333	1.589	39	1.599	25
511				
520			1.546	3
432				
521			1.519	6
440	1.460	74	1.469	55

The exchange reaction can be written:

$$(Co_2Li_2)V_2O_8 + CoSO_4 \xrightarrow{350°C} (Co_3/□)V_2O_8 + Li_2SO_4 .$$

disordered spinel   ordered spinel

The new variety has a spinel structure, with ordered distribution of vacancies and Co^{2+} over the octahedral sites in the ratio $\frac{1}{3}$. This form is isostructural with the lacunary spinels $(T_3^{4+} \square)M_2^{2+}O_8$ described in Section 3.2. Table 4.9 and Fig. 4.1 give the X-ray patterns of the lithium containing spinel and of the exchanged compound.

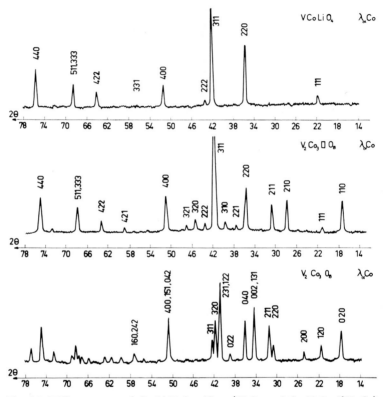

Fig. 4.1. Diffractograms of $Co_2Li_2V_2O_8$, $(Co_3 \square)V_2O_8$ and $Co_3V_2O_8$ ($\lambda K\alpha Co$).

Remark: we emphasize that starting with a disordered spinel, we got an exchanged compound with ordered octahedral sites, at a temperature as low as 350°C.

It appears that the cation diffusion is tremendously increased by the vacancies and consequently the order is established in a rather short time.

a) Thermal stability of the defect form of cobalt orthovanadate
The DTA curve shows two thermic accidents (Fig. 4.2). The first one endo-

thermic, at 520°C, corresponds, probably to disordering of vacancies and cations on the octaedral sites. The second one at 570 °C, more important and exothermic, corresponds to the transition to the stable orthorhombic form, according to the irreversible reaction:

$$(Co_3/\square)V_2O_8 \xrightarrow{\;570\,°C\;} Co_3V_2O_8 \;.$$
$$\underset{\text{cubic metastable form}}{} \qquad \underset{\text{orthorhombic stable form}}{}$$

570°C is the temperature of rapid transformation, determined by using a heating rate of 300°C/hour. As a matter of fact, if one keeps a sample of the defect cubic form at 460°C – that is to say 100°C below the temperature of rapid transformation – one finds after 3 or 4 weeks, about 50 per cent of the starting material changed into the orthorhombic form. By heating the sample at an even lower temperature, one would get the same result, but only after a much longer period.

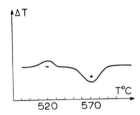

Fig. 4.2. DTA curve of ($Co_3\,\square$)V_2O_8 compound (heating speed 300°C/hour).

Table 4.10

Cell parameters of cobalt orthovanadates

Formula	Cell parameter (Å)	Cell volume U (Å^3)	Density d_x (g/cm^3)
$VCoLiO_4$	$a = 8.276 \pm 0.003$	567	4.24
$V_2Co_3\,\square\,O_8$	$a = 8.324 \pm 0.005$	577	4.68
$V_2Co_3O_8$	$\begin{cases} a = 8.310 \\ b = 6.03 \approx a/\sqrt{2} \\ c = 11.48 \approx a\sqrt{2} \end{cases}$	575	4.69

It is likely that the electrostatic energy gain due to the vacancy order on the octahedral sites plays an important part in the stability of the metastable phase.

References p. 208

b) Comparison between the main crystalline features of the two structural forms of cobalt orthovanadate

Parameters: Cell parameters, volumes and densities are listed in Table 4.10 for the starting lithium cobalt vanadate and the two allotropic forms.

Anion sublattice: In the two allotropic phases, the oxygen anions form a cubic close packing. As the cations radius are rather small ($V^{5+}=0.59$ Å; $Co^{2+}=0.72$ Å). The cell parameters are imposed by the oxygen stacking, which is very little affected by the transition and so are the calculated densities (Table 4.10). *Cation distribution*: Table 4.11 gives the cation distribution in the 2 space groups.

Table 4.11

Ion distribution in the two allotropic forms of cobalt orthovanadate

---

a) $V_2Co_3 \square O_8$. Metastable form
Group P4$_3$32 (or P4$_1$32)

8 V^{5+}	in	8 (c)	$x \approx 0$
12 Co^{2+}	in	12 (d)	$x \approx \frac{3}{8}$
4 $\square$	in	4 (b)	
24 O^{--}	in	24 (e)	$x = y \approx \frac{1}{8}, z \approx \frac{3}{8}$
8 O^{--}	in	8 (c)	$x \approx \frac{3}{8}$

b) $V_2Co_3O_8$. High temperature form
Group Abam

4 Co^{2+}	in	4 (a)	
8 Co^{2+}	in	8 (c)	$y \approx \frac{1}{8}$
8 V^{5+}	in	8 (f)	$x \approx \frac{1}{8}; y \approx \frac{3}{8}$
8 O	in	8 (f)	$x \approx \frac{1}{2}; y \approx \frac{1}{2}$
8 O	in	8 (f)	$x \approx \frac{1}{2}; y \approx 0$
16 O	in	16 (g)	$x \approx 0; y \approx \frac{1}{8}; z \approx \frac{1}{2}$

Fig. 4.3 shows a projection of the two cation positions on the (001) planes. One can see an important difference. In the orthorhombic phase, the distance between the centers of non occupied octahedra or tetrahedra (of the oxygen close packing) to the nearest occupied tetrahedron (V^{5+} site) is less than 2 Å. Now the minimum cation–cation distance consistent with electrostatic stability of an oxide compound is about 2.80 Å, thus this structure is "vacancy-free" and unable to accept extra cations.

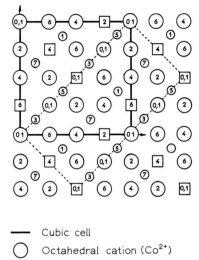

—   Cubic cell

◯   Octahedral cation (Co^{2+})

▢   Vacancy

○   Tetrahedral cation (V^{5+})

Fig. 4.3 (a). Structure of the metastable form of cobalt orthovanadate: projection of cations on (001) plane (numbers in eighths of c parameter).

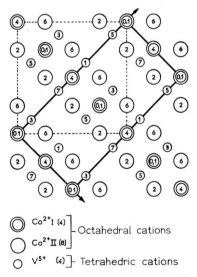

◎ $Co^{2+}I$ (4) $\Big]$ — Octahedral cations
◯ $Co^{2+}II$ (8) $\Big]$

○ V^{5+} (4) $\Big]$ — Tetrahedric cations

Fig. 4.3 (b). Structure of the high temperature form of cobalt orthovanadate: projection of the cations on (001) plane (numbers in eighths of c parameter).

References p. 208

On the other hand, in the metastable form, the so called "defect" positions could very well be occupied by cations: the $\square - Co^{2+}$ distances are more than 2.9 Å and the $\square - V^{5+}$ distances more than 3.4 Å.

This example establishes a clear distinction between what we will call "lacunary" metastable allotropic phase and the stable "vacancy-free" phase.

New allotropic metastable forms of cobalt and magnesium orthophosphate

a) Structure of the high temperature forms of cobalt and magnesium ortho-phosphates $Co_3P_2O_8$ and $Mg_3P_2O_8$

These two compounds are isostructural. The structure was investigated by Calvo [1967]. The symmetry is monoclinic and belongs to the space group $P2_1/C$. The cell parameters are (Å):

$Co_3P_2O_8$	$Mg_3P_2O_8$
$a = 7.557$	$a = 7.01$
$b = 8.365$	$b = 8.21$
$c = 5.067$	$c = 5.03$
$\beta = 94°05$	$\beta = 90°25$.

The elementary cell contains 2 formula units ($Z = 2$).

Among the six Co^{2+} (Mg^{2+}) distributed over the unit cell, four have a tetrahedral and two an octahedral oxygen environment. The PO_4 tetrahedra do not share corners. The oxygen packing is thus rather loose.

By solid state exchange reaction, we succeeded in preparing a new metastable allotropic form of these two orthophosphates.

The exchange reactions are written below:

$$(Li_2/Mg_2)P_2O_8 + MgSO_4 \xrightarrow{660°C} (Mg_3/\square)P_2O_8 + Li_2SO_4$$

$$(Li_2/Co_2)P_2O_8 + CoSO_4 \xrightarrow{680°C} (Co_3/\square)P_2O_8 + Li_2SO_4.$$

b) Starting materials

The lithium containing starting materials $(Mg_2/Li_2)P_2O_8$ and $(Co_2/Li_2)P_2O_8$ belong to the olivine group (Newnham et al. [1965]). The orthorhombic space group is $Pmnb$. The cell parameters are given in Table 4.12. The oxygen packing is hexagonal compact. There are two kinds of octahedral sites in the structure respectively occupied by Li^+ and Mg^{2+} (Co^{2+}).

c) The defect exchanged phases
By analogy with the starting materials' cells, it was easy to determine those of the new compounds; the values are listed in Table 4.12 along with those corresponding to the lithium compounds. Figs. 4.4 and 4.5 show the analogy between the X-ray diffractograms of the starting and the exchanged materials. Of course, due to the structural differences, no analogy exists between the diffractograms corresponding to the two allotropic forms of the same compounds.

Table 4.12

Cells parameters of olivine orthophosphates and corresponding metastable exchanged compounds (Å)

	$(Mg_2/Li_2)P_2O_8$	$(Mg_3/\square)P_2O_8$	$(Co_2/Li_2)P_2O_8$	$(Co_3/\square)P_2O_8$
a	5.87	5.911	5.92	5.920
b	10.15	10.214	10.20	10.334
c	4.68	4.734	4.70	4.75
γ		90°60		91°04

The structure is very close to that of the starting material: half of the lithium sites are occupied by new bivalent cations. The symmetry cannot be orthorhombic anymore, however, the monoclinic distortion is very weak. Writing the full group symbol of the starting material $P(2_1/m\,(2_1/n)\,2_1/b)$, the monoclinic space group of the defect compound could be $P2_1/m$, $P2_1/n$ or $P2_1/b$, with three different vacancy orderings.

Systematic X-ray extinctions show the space group to be $P2_1/b$, so that in the elementary cell, the vacancies are located on the symmetry centers

Fig. 4.6. represents the projection of the cations on the (001) plane, in the defect form of cobalt orthophosphate. Starting with the available values for the position parameters of the original olivine materials, we refined the position parameters corresponding to the new lacunary compounds. The results are listed in Table 4.13. Tables 4.14 and 4.15 reproduce the recording of the new lacunary olivine compounds.

d) Thermal stability of the two defect olivine phases
The metastable compounds transform into the stable vacancy-free forms,

References p. 208

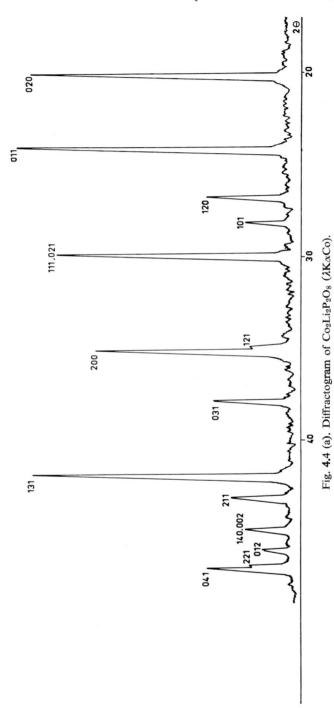

Fig. 4.4 (a). Diffractogram of $Co_2Li_2P_2O_8$ ($\lambda K\alpha Co$).

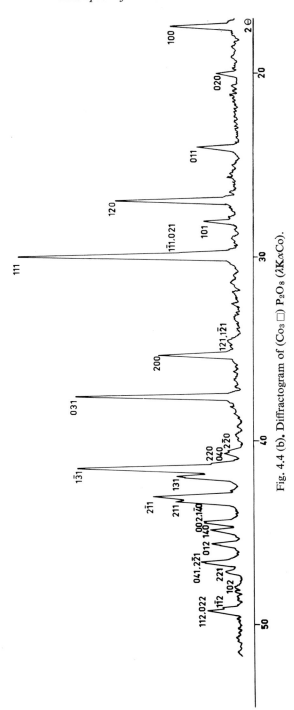

Fig. 4.4 (b). Diffractogram of $(Co_3 \square) P_2O_8$ $(\lambda K\alpha Co)$.

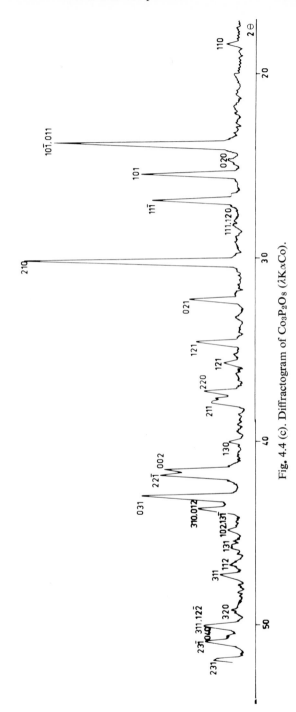

Diffractogram $Co_3(PO_4)_2$

Fig. 4.4 (c). Diffractogram of $Co_3P_2O_8$ ($\lambda K\alpha Co$).

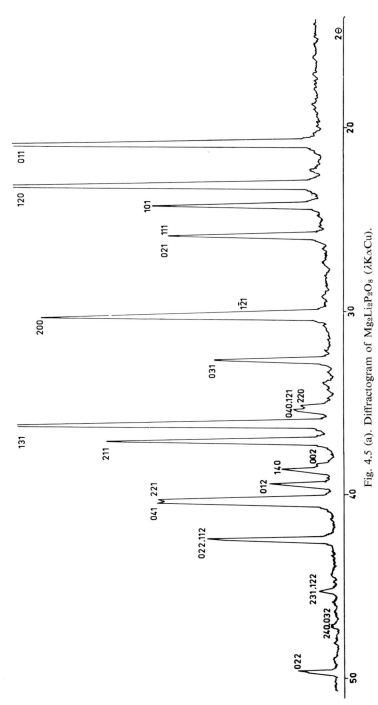

Fig. 4.5 (a). Diffractogram of $Mg_2Li_2P_2O_8$ ($\lambda K\alpha Cu$).

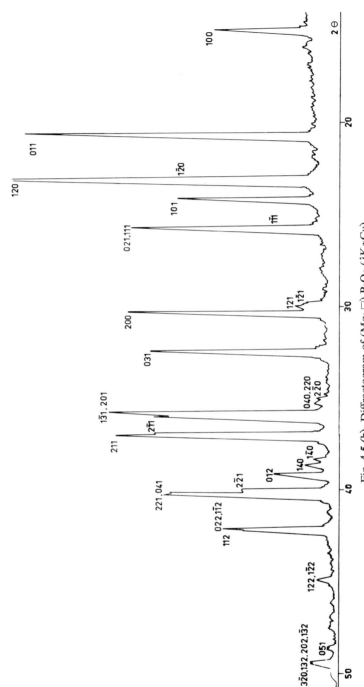

Fig. 4.5 (b). Diffractogram of $(Mg_3 \square) P_2O_8$ ($\lambda K\alpha Cu$).

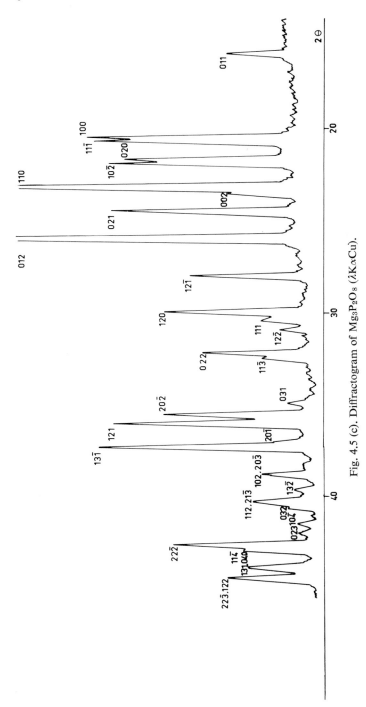

Fig. 4.5 (c). Diffractogram of $Mg_3P_2O_8$ ($\lambda K\alpha Cu$).

according to the two weakly exothermic reactions:

$$(Mg_3 \square)P_2O_8 \xrightarrow{820\,°C} Mg_3P_2O_8$$

$$(Co_3 \square)P_2O_8 \xrightarrow{870\,°C} Co_3P_2O_8 .$$

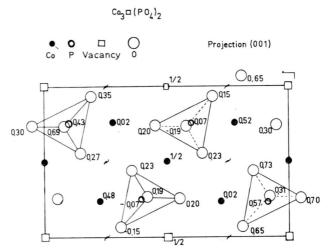

Fig. 4.6. Structure of $(Co_3 \square)P_2O_8$: projection of the cations on the (001) plane.

Table 4.13

Position parameters of the ions in $(Mg_3 \square)P_2O_8$ and $(Co_3 \square)P_2O_8$ structures

Position group $P2_1/b$	$(Mg_3 \square)P_2O_8$			$(Co_3 \square)P_2O_8$		
	x	y	z	x	y	z
$2\square$ in 2 (a)						
$2M_{II}^{2+}$ in 2 (b)						
$4M_{I}^{2+}$ in 4 (e)	0.236;	0.277;	−0.023	0.227;	0.278;	−0.017
4P in 4 (e)	0.242;	0.103;	0.426	0.277;	0.100;	0.430
$4O_I$ in 4 (e)	0.266;	0.079;	0.306	0.279;	0.081;	−0.258
$4O_{II}$ in 4 (e)	0.259;	0.453;	0.202	0.246;	0.462;	0.193
$4O_{III}$ in 4 (e)	0.059;	0.190;	0.350	0.096;	0.168;	0.298
$4O_{IV}$ in 4 (e)	0.439;	0.145;	0.272	0.493;	0.139;	0.258

Table 4.14

Recording of $(Mg_3\ \square)P_2O_8$ ($\lambda K\alpha Cu$)

hkl	$d_{obs.}$	$I_{obs.}$	hkl	$d_{obs.}$	$I_{obs.}$
100	5.896	12	1̄12 ⎫		
020		0	022 ⎬	2.1486	142
011	4.293	64	112 ⎪		
1̄20 ⎫	3.881	151	2̄31 ⎭		
120 ⎭	3.838		1̄22	2.022	32
101	3.691	32	122 ⎭		
1̄11 ⎫			051	1.8730	28
021 ⎬	3.487	75	1̄32 ⎫	1.8478	
111 ⎭			202 ⎪	1.8478	36
1̄21 ⎫	3.004		3̄20 ⎪		
121 ⎬	2.983	141	132 ⎭		
200 ⎭	2.950		320		
031	2.761	112	2̄12		0
2̄20 ⎫	2.566		301		0
040 ⎬		57	212 ⎫		0
220 ⎭	2.548		2̄41 ⎪		
1̄31 ⎫			3̄11 ⎪		
201 ⎬	2.513		⎬	1.794	
⎭		301	151 ⎪		135
131	2.490		311 ⎭		
2̄11	2.439	280	151 ⎭	1.7822	
211 ⎭	2.427		241		
002		4	2̄22 ⎫	1.7417	
1̄40	2.352	14	042 ⎪		432
140	2.332	16	222 ⎪		
012	2.306	61	3̄21 ⎬		21
2̄21 ⎫	2.258		321 ⎪		
041 ⎬	2.2476	268	060 ⎭	1.7070	87
221 ⎭	2.2410				

The rapid transformation temperatures were determined by D.T.A. (heating speed 300°C/h).

Here again the reaction seems to be irreversible. In both cases, the transformation produces an increase of the cell volume of about 10%. Thus in spite of the presence vacancies, the defect forms have a higher density than the stable high temperature forms, in which the oxygen packing is no more compact.

References p. 208

Table 4.15

Recording of $(Co_3 \square)P_2O_8$ ($\lambda K\alpha Co$)

hkl	$d_{obs.}$	$I_{obs.}$	hkl	$d_{obs.}$	$I_{obs.}$
100	5.947	50	102	2.203	27
020	5.177	18	1$\bar{1}$2	2.157⎫	
011	4.32	58	022		227
1$\bar{2}$0		≈ 0	112	2.151⎭	
120	3.86	266	1$\bar{4}$1		≈ 0
101⎫	3.710⎫	38	141		≈ 0
1$\bar{1}$1⎪	3.496⎪		2$\bar{3}$1		≈ 0
021⎪		504	1$\bar{2}$2	2.030⎫	171
111⎭	3.478⎭		122	⎭	
1$\bar{2}$1	3.028	4	231		≈ 0
200	2.958	227	300	1.972⎫	45
031	2.790	547	2$\bar{4}$0	⎭	
2$\bar{2}$0	2.587⎫		240		≈ 0
040	2.581⎪	97	051		89
220	2.548⎬		3$\bar{2}$0		
1$\bar{3}$1	2.536⎪	1172	202		167
202	⎭		132	1.8508⎭	
131	2.507⎫	790	320		≈ 0
2$\bar{1}$1	2.448⎪		2$\bar{1}$2		
1$\bar{4}$0	2.382⎬		301	1.821	137
002	2.372⎪	271	212		
140	2.352⎭		3$\bar{1}$1	1.798⎬	
012	2.313	160	151	1.794⎪	898
2$\bar{2}$1	2.273⎫		311		
041	2.270⎭	374	241	⎭	
			2$\bar{2}$2		1662

Remark: We note again that in these two examples, the oxygen network is not affected by the exchange process, and that the structure of the lacunary phase remains still very close to that one of original lithium compound.

4. Conclusion

In this section, we described a new method for the synthesis of metastable defect oxide compounds. Starting with a lithium containing material, we performed the exchange:

$$2\text{Li}^+ \rightarrow \text{M}^{2+} + \square$$

by solid state reaction.

Several new metastable defect spinel and olivin materials with ordered vacancies were prepared by this method, namely:

$$\left.\begin{array}{l}(\text{Ti}_3^{4+}\square)\text{M}_2^{2+}\text{O}_8 \\ (\text{Ge}_3^{4+}\square)\text{M}_2^{2+}\text{O}_8\end{array}\right\} \text{ with } \text{M}^{2+} = \text{Zn, Co, Mg, Cd, Mn}$$

$$(\text{Mn}_3^{4+}\square)\text{Zn}_2^{2+}\text{O}_8$$

$$(\text{Co}_3^{2+}\square)\text{V}_2^{5+}\text{O}_8$$

$$(\text{Ti}_4\text{Co}\square)\text{Co}_3\text{O}_{12}$$

Defect spinel compounds

$$\left.\begin{array}{l}(\text{Mg}_3\,\square)\text{P}_2^{5+}\text{O}_8 \\ (\text{Co}_3\square)\text{P}_2^{5+}\text{O}_8\end{array}\right\} \text{ Defect olivine compounds.}$$

All these new defect materials collapse when heated at rather low temperatures (a few hundred degrees C).

From these positive tests, we are in a position to derive some conclusions concerning the exchange process and the metastable character of the defect compounds.

The solid state exchange reaction in lithium containing oxide compounds $2\text{Li}^+ \rightarrow \text{M}^{2+} + \square$ has a topotactical character in as much as the exchanged compounds preserve the essential features of the original structures. The presence of vacancies enhances cation diffusion and gives rise to well vacancy ordered structures which belong to subgroups of the original groups.

Heating at high enough temperatures produces in a first stage vacancy disorder and in a second stage the collapse of the structure either into its oxidic constituents or to a vacancy-free allotropic form. The irreversible nature of this high temperature reaction definitely classifies the vacancy ordered structures as metastable. As there is no subgroup correlation between the vacancy ordered and the vacancy-free allotropic form, the irreversible transition from the first to the second one is first order.

Conditions for the success of the topotactical exchange reaction are a stable anion network, a comparable size of the substituted ions and a favorable heat of formation difference between the exchange agent (M^{2+}SO_4) and the reaction product (Li_2SO_4).

References p. 208

Further solid state exchange reactions can be generalized in a scheme like
$$n\mathrm{A}^{m+} \to m\mathrm{B}^{n+} + (n-m) \square \quad (n > m).$$
We are quite confident that they are possible for families other than oxides and with various choices of n and m.

References

BERTHET, G., 1968, Thèse 3e Cycle, Contribution à l'étude des composés lacunaires ioniques (Grenoble).

CALVO, C., 1967, Colloque International sur les phosphates minéraux solides (Toulouse, France).

DURIF, A. and J. C. JOUBERT, 1962, C. R. Acad. Sci. (Paris) **255**, 2471–2473.

JOUBERT, J. C., 1965, Thèse d'Etat, Contribution à la recherche et à l'étude des surstructures dans les oxydes mixtes du type spinelle. Préparation de composés spinelles lacunaires (Grenoble).

JOUBERT, J. C. and A. DURIF, 1964, C. R. Acad. Sci. (Paris) **258**, 4482–4485.

NEWNHAM, R. and M. J. REDMAN, 1965, J. Amer. Ceram. Soc. **48**, 547.

WYART, J. and G. SABATIER, 1956 a, Bull. Soc. Fr. Miner. Crist. **79**, 444–448.

WYART, J. and G. SABATIER, 1956 b, Bull. Soc. Fr. Miner. Crist. **79**, 574–581.

CHAPTER 5

ON STOICHIOMETRY IN TWO-DIMENSIONAL CHEMISORBED PHASES

J. BÉNARD

Ecole Nationale Supérieure de Chimie de Paris, France

1. Introduction

In the last few years the simultaneous use of two new techniques – the measurement of adsorption on single crystals by means of radioisotopes and the diffraction of low-energy electrons – has led to a better understanding of the structure and properties of adsorbed layers. Research has been carried out, in particular, in our laboratory on the adsorption of sulphur on various metals by these two methods.

It appears that in these metal-sulphur systems, there form, on surfaces with low crystallographic indices, genuine two-dimensional compounds with essentially ionic character. The study of the stoichiometry and the distribution of different species of ions in these compounds leads to some interesting observations which may be compared with the known properties of three-dimensional ionic compounds.

The experimental data to which we will refer have been published for the most part, in recent years. I will review them briefly; the details will be found in the original publications. However, I shall examine in more detail the problems connected with the stoichiometry of adsorbed compounds which come directly into the framework of the present volume.

2. Stoichiometry of adsorbed layers

Experimental work carried out over a period of about ten years in our laboratory has led to a very accurate method for determining adsorption isotherms of sulphur on single crystals of different metals in conditions of s'rict reversibility (Bénard [1960]). The method consists in treating a surface

References p. 218

of low crystallographic indices, free of impurities, in a mixture of hydrogen and hydrogen sulphide with a composition such that the system is reversible at the temperature of the experiment. By using hydrogen sulphide marked with ^{35}S and by means of a carefully defined experimental procedure it is possible to measure, after the treatment, the quantity of sulphur chemisorbed at equilibrium. Knowing the real area of the plane surface that has been subjected to adsorption one then determines the surface concentration of the adsorbed atoms. From a series of experiments in gaseous media with different values of the ratio pH_2S/pH_2 one obtains a curve: surface concentration against composition of the gas mixture, which is an adsorption isotherm.

These isotherms have the following characteristics:

1. Obtained in strictly reversible conditions without hysteresis, they correspond to an equilibrium state of the chemical system comprising the three phases: gaseous phase, solid phase and adsorbed "phase". They thus enable one to define the thermodynamic properties and in particular the free enthalpy, the isosteric enthalpy and, to a certain degree, the entropy.

2. As the isotherms are for plane monocrystalline surfaces, for which the perfection has been tested by various methods, they define an adsorption process on a well defined crystallographic substrate.

3. The isotherms are obtained in conditions of temperature and concentration such that no three-dimensional sulphide can be formed. One can thus be sure of attributing the observations to adsorption with the exclusion of any classical reaction of sulphide formation.

Thus, all the experimental conditions are chosen at the outset to lead to a study of the relation between the thermodynamic and the structural aspects of adsorption. Let us recall here briefly the references to the experimental work. Oudar [1959] determined for the first time a reversible adsorption isotherm of sulphur on polycrystalline copper. Cabané-Brouty and Oudar [1964] (Cabané-Brouty [1965]) then traced the reversible adsorption isotherms of sulphur on the principal faces of silver. More recently Perdereau [1969] has determined a part of the isotherms for the principal faces of nickel; the complete determination for this system comes up against insurmountable experimental difficulties. We reproduce, as examples, the isotherms obtained at 400°C on the (111) and (110) faces of silver. The results (Fig. 5.1) are plotted in semi-logarithmic coordinates in order to develop the beginning of the coverage of the surface by sulphur atoms. On the abscissa is plotted the logarithm of the ratio pH_2S/pH_2 which defines the composition of the gas phase. The surface concentration of sulphur is plotted on the ordinate in units of mass or number of atoms of adsorbed sulphur per unit surface area. We shall make some remarks on these isotherms which can, in fact, be applied to all the isotherms obtained in this research.

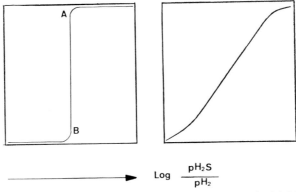

Fig. 5.1. Adsorption isotherms of sulphur on the (111) [left] and (110) [right] faces of silver (Cabané-Brouty).

As the partial pressure of hydrogen sulphide in the gas phase increases, the surface concentration of adsorbed atoms increase from a value which is practically zero in pure hydrogen, to a maximum value which corresponds to a saturation plateau. The variation can be very gradual, as in the case of the (110) face of silver, or very sharp, as for the (111) face.

In each case the shapes of the isotherms indicate, by comparison with the Langmuir isotherm, that there exist, over a wide range of concentration, attractive forces between adsorbed particles.

The surface concentration of sulphur on the plateaux of the isotherms is obtained with good precision and is perfectly reproducible. The value is always lower than that which would correspond to a monoatomic layer of doubly ionized sulphur atoms in a compact arrangement on the metal surface. The difference is marked and cannot be attributed to experimental error. Moreover, the value always has the same order of magnitude not only for the different crystalline faces of the same metal but also for the different metals studied. The value is independent of temperature (Bénard et al. [1965]). In Table 5.1, the concentrations at saturation measured on the

References p. 218

different crystalline faces of copper, silver, nickel and α-iron are presented in $(g/cm^2) \times 10^9$.

The average value of the figures in this table differs by about 30 per cent from the value corresponding to a compact arrangement of doubly ionized

Table 5.1

Copper			Silver		
(111)	(100)	(110)	(111)	(100)	(110)
39 ± 2	37.5 ± 2	37 ± 2	39 ± 1	38 ± 1	36 ± 1
Nickel			α-iron		
(111)	(100)	(110)	(110)	(211)	(100)
47 ± 1	43 ± 1	—	45.6 ± 1	39.6 ± 1	35.4 ± 1

sulphur. This arrangement may therefore be excluded together with any arrangement of uniform distribution of sulphur atoms on sites of high coordination – which would lead, as we shall see later, to surface concentrations much lower than those observed. This type of arrangement would result, moreover, in differences in concentrations between the different crystalline orientations which would be much greater than those actually observed.

A careful examination of the figures in Table 5.1 shows that although the orders of magnitude are the same there appear small but significant and reproducible differences. In particular, it is always on the surfaces with the highest density of substrate atoms that the concentration of adsorbed sulphur is greatest: (111) for copper, silver and nickel, (110) for α-iron.

The saturated surface concentration of adsorbed atoms, or in other words *the stoichiometry of the adsorbed layer*, is thus controlled by factors other than the principle of maximum utilization of the available space on a monocrystalline surface. But equally excluded are factors connected with the geometry of crystallographic sites on the metal surface, which is much less in agreement with the hypotheses frequently advanced in chemisorption. Only a crystallographic study of the distribution of adsorbed atoms by diffraction of low-energy electrons, a technique that has become fully operational only in recent years, could lead to a clarification of these points.

3. Structure of adsorbed layers

The adsorbed monolayers for which the surface concentrations had been measured as indicated above were examined by the diffraction of electrons with low energies (of the order 100 eV).

As it is impossible to reproduce in the diffraction apparatus (Varian) the gaseous conditions used for determining the isotherms in conditions of strict reversibility, we had to arrive at the saturated states by progressive exposure of the monocrystalline surfaces to traces of hydrogen sulphide. The surfaces were previously cleaned by repeated bombardment with argon ions and heating in vacuo of 10^{-10} torr. Although it is quite evident that these conditions are very far from those of reversibility, it was easy to establish, at least for the saturated states, the correspondance between the observations made in these two series of experiments.

I will not go into detail here on the difficulties encountered in obtaining the diffraction patterns and in their interpretation. I will recall only that the interpretation has to be limited, at the moment, to the identification of the symmetry of the surface layers; in the absence of a complete theory of the diffraction of electrons at these low energies any interpretation of intensities has not been possible.

The analysis of the diffraction patterns for the saturated states, that is the states corresponding to the plateaux of the adsorption isotherms, leads to the conclusion that, whatever the metal and whatever the crystalline face, the distribution of surface atoms in these states differs from that at the clean metal surface.

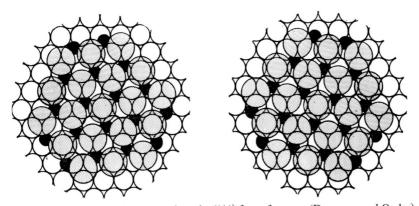

Fig. 5.2. Two-dimensional compound on the (111) face of copper (Domange and Oudar).

References p. 218

Considering this fact, together with the observation that there is appreciable space between the adsorbed sulphur atoms, we are led to adopt the hypothesis first put forward by Farnsworth and Madden [1961] and by McRae [1964] for the case of metal oxygen systems: the adsorbed layer is made of two kinds of atoms: metal and non-metal, the ionized metallic atoms occupying the empty spaces of the two-dimensional lattice. The adsorbed layer at saturation can thus be considered as a *true two-dimensional compound* with a structure different from that of the substrate and which is related to that of the substrate by well defined epitaxial relations.

As an example Fig. 5.2 shows the structure of the compound formed on copper (111) which possesses two possible, symmetrical, positions with respect to the substrate (Domange and Oudar [1967b]). For comparison Fig. 5.3 shows the structure of the compound formed on the (100) face of the same metal (Domange and Oudar [1967a]). Particularly interesting in this respect is the (111) face of nickel which at saturation is covered by a surface compound with quaternary symmetry although the arrangement of the substrate atoms corresponds to ternary symmetry.

Numerous structure analysis have also been made of the *incompletely covered surfaces*. The intermediate stages in the building up of the saturated layer are sometimes complex but, whatever the system, one always passes through a preliminary stage which corresponds to the fixing of sulphur atoms at sites of high coordination of the substrate surface. These primary structures evidently possess the same symmetry as the substrate, but with different periodicities. One can consider them to a certain degree, as surface

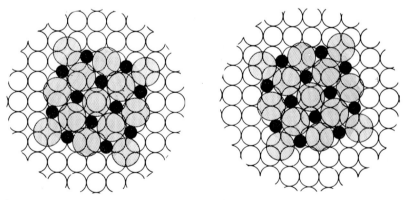

Fig. 5.3. Two-dimensional compound on the (100) face of copper (Domange and Oudar).

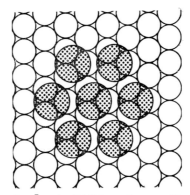

Fig. 5.4. Precursor phase on the (111) face of copper (Domange and Oudar).

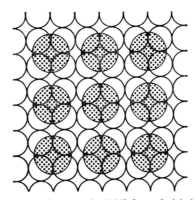

Fig. 5.5. Precursor phase on the (100) face of nickel (Perdereau).

compounds, although they differ from the previous ones, which to a first approximation have an unchangeable structure and develop by a process of nucleation and lateral growth, whereas these structures grow by progressive and random filling of available sites on the substrate surface. Fig. 5.4, for example, shows schematically the precursor state observed on copper (111) and Fig. 5.5 the precursor state observed on nickel (100) (Perdereau [1968a, 1968b]).

The transition of the precursor states to the states of maximum saturation, following the simultaneous increase of the concentration of sulphur and the temperature, sometimes brings into play complex processes, which we will not deal with here.

References p. 218

4. Discussion

If we limit discussion to the structural data that we have just outlined we may consider, as a first approximation, that the *surface compounds observed at saturation are two-dimensional stoichiometric phases, while the precursor states in which the atoms are localized on special sites of the substrate are nonstoichiometric phases.*

It is not without interest to note that the "compositions" of these phases and their variations depend essentially on geometrical factors, in contrast to what takes place in three-dimensional ionic phases where the necessity of maintaining the electrostatic neutrality of the crystalline lattice imposes *a priori* conditions, independently of the stereochemical factors. This situation arises from the fact that the atoms of the adsorbed layer exchange bonds not only between themselves in the plane of the layer but also with the atoms of the metallic substrate.

From the point of view of the concentration of the adsorbate in these two types of surface phases, the passage from one to the other brings about a sharp increase in concentration and a simultaneous decrease in the minimum

Table 5.2

	Copper (111)		Copper (100)	
	Precursor state	Saturated phase	Precursor state	Saturated phase
Ratio S/Me	3/9	3/7	8/32	8/17
Distance S–S	4.42 Å	3.90 Å	5.1 Å	3.72 Å

sulphur-sulphur spacing. For example, in the case of copper the ratio of the number of sulphur atoms to the number of metal atoms in the surface (S/Me) varies, for the (111) and (100) faces, as shown in Table 5.2.

It is interesting to note that the minimum distance between sulphur atoms in the copper (111) surface compound, equal to 3.90 Å, is very close to that of sulphur atoms in the (111) plane of three-dimensional cuprous sulphide, Cu_2S, (3.94 Å).

The problem of the relationship between the variations of surface concentration and the structure of the surface phases presents a very special aspect in the case of the (110) face of copper, as have shown Domange and Oudar [1968]. In the clean perfect state this face is made up of parallel atomic grooves with separation 3.61 Å. The first atoms of sulphur fix themselves along these grooves on substrate sites with the minimum separation of 5.1 Å imposed in this direction (Fig. 5.6). This precursor state, which is filled progressively, is comparable *mutatis mutandis* with the precursor states

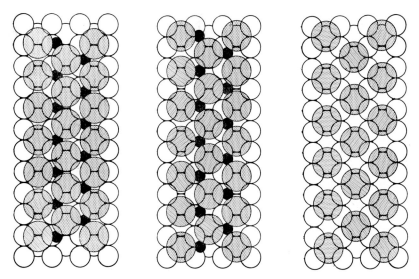

Fig. 5.6. Series of two-dimensional structures on the (110) face of copper (Domange and Oudar).

observed on other faces; it is quickly succeeded by a more condensed state characterized by a minimum S–S distance, along the rows, of 4.26 Å. At this point, if the pressure of sulphur is increased, there occurs a contraction which is regular, and no longer discontinuous, along the direction of the rows which themselves remain at the same spacing. The sulphur–sulphur spacing thus changes progressively from 4.26 to 3.85 Å. It thus appears that for the case of copper (110) *the phase corresponding to complete saturation of the surface is a true two-dimensional nonstoichiometric phase* in which the variations of composition are accompanied by the variation of a single parameter of the crystalline lattice.

It is evident that, at this stage of the research, it would be profitable to establish a more strict parallel between the structural observations and the shape of the adsorption isotherms which depend, as we have seen, on the crystalline orientation. Unfortunately, the experimental conditions in which

References p. 218

the isotherms are obtained are usually different from those in which the diffraction patterns are observed. Any detailed analysis of the problem would thus be premature at the moment. At the best one can affirm that the steep isotherms such as that observed on silver (111) (Fig. 5.1) must characterize systems in which strictly stoichiometric two-dimensional surface compounds formed by nucleation and growth, whereas the isotherms which climb more gradually, such as that observed on (110), must characterize systems in which the structure, and in consequence the composition, of the surface phases are less strictly defined.

To conclude, it seems interesting to compare these studies with our knowledge of the formation of insertion compounds in lamellar lattices: graphite, montmorillonites, and MoS_2 type sulphides. The formation of metal-sulphur surface phases, in effect, satisfies both thermodynamic and stereochemical requirements for which equivalents are easily found in the formation of threedimensional interstitial phases. Such a comparison might provide a general view of the problem of stoichiometry which would include both adsorption phenomena as well as the variations of composition in conventional three-dimensional compounds.

I would like to thank in particular J. Oudar, J. L. Domange and M. Perdereau for fruitful discussions and G. Rhead for the translation into English.

References

BÉNARD, J., 1960, Bull. Soc. Chim. Fr., p. 203.
BÉNARD, J., 1960, Proc. XVIIe Congrès Union Intern. Chimie (Butterworth) p. 109.
BÉNARD, J., J. OUDAR and F. CABANE-BROUTY, 1959, Surface Science 3, 359.
CABANÉ-BROUTY, F., 1965, J. Chim. Phys. Fran. 62, 1045 and 1056.
CABANÉ-BROUTY, F. and J. OUDAR, 1964, C. R. Acad. Sc. 258, 5428.
DOMANGE, J. L. and J. OUDAR, 1967a, C. R. Acad. Sc. 264, 35.
DOMANGE, J. L. and J. OUDAR, 1967b, C. R. Acad. Sc. 264, 951.
DOMANGE, J. L. and J. OUDAR, 1968, Surface Science 11, 124.
FARNSWORTH, H. E. and H. H. MADDEN, 1961, J. Appl. Phys. 32, 1933.
MacRAE, A. U., 1964, Surface Science 1, 319.
OUDAR, J., 1959, C. R. Acad. Sc. 249, 91.
PERDEREAU, M., 1968a, C. R. Acad. Sc. 267, 1107.
PERDEREAU, M., 1968b, C. R. Acad. Sc. 267, 1288.
PERDEREAU, M., 1969, Thèse Paris.

CHAPTER 6

NONSTOICHIOMETRY AND SINTERING IN IONIC SOLIDS

P. J. L. REIJNEN

Philips Research Laboratories, Eindhoven, Netherlands*

1. Introduction

In this chapter the influence of nonstoichiometry on the complex sintering process of ionic solids is discussed. During sintering of ionic solids both the cations as well as the anions must be transported by volume, surface and grain boundary diffusion. It is shown that small deviations from stoichiometry can have a marked influence on volume diffusion during sintering. The dependence of surface and grain boundary diffusion on small deviations from stoichiometry is difficult to predict, but volume diffusion can quite well be correlated theoretically with a given model for the defect structure of the solid. This is done in the theoretical part of this article. The results of the calculations are compared with sintering experiments on simple oxide systems which have been given a controlled deviation from stoichio-metry. The agreement between theory and experiment is quite good, indi-cating that volume diffusion is the predominant diffusion path, at least in these sintering experiments. The effect of nonstoichiometry on sintering has been considered previously by many other authors. In most papers published about fifteen years ago one finds the reflection of the idea that the total number of vacancies, both anion and cation vacancies, determines volume diffusion; see e.g. Marshall et al. [1954] and Jones et al. [1958]. However, Jorgensen [1965] pointed out that if volume diffusion of Al^{+++} during sintering of alumina is rate-determining for the overall process, there will be a decrease of the sintering rate when MgO is dissolved in alumina. The oxygen vacancy concentration is increased upon dissolution of MgO, but the cation vacancy concentration is decreased and consequently diffusion of the rate-determining species is further hampered. Readey [1966] in a paper entitled

---

* Now with R.T.C., 41 Rue Pierre Brossolette, 27 Evreux, France.

References p. 238

"Mass Transport and Sintering in Impure Ionic Solids" has calculated the dependence of volume diffusion on small deviations from stoichiometry.

Readey's work and the present paper are in agreement in their overlapping parts.

Johnson [1967], however, comes to just the opposite conclusion, but writes that the result of his calculations are "unexpected and unobserved". In the theoretical part of this chapter the work of Johnson will be commented on. In the discussion at the end of the chapter an attempt is made to unravel the complex relation between grain growth, sintering rate, dispersed second phase, deviation from stoichiometry and discontinuous grain growth.

2. Theory

2.1. The origin of compositional gradients in sintering solids

The mobility of ions in a solid is made possible by the presence of point defects in the lattice. Point defects are deviations from the ideal lattice, such as unoccupied normal sites (vacancies) or ions at sites which are normally not occupied (interstitial ions). An ion at a normal site may jump into a vacancy, leaving a vacancy behind. Other possibilities for the motion of ions are a jump of an ion into an interstitial position or a jump of an interstitial ion into another interstitial position. The following notation is used in this paper to denote point defects in a solid MO:

V_c cation vacancy which is formed when a M^{++} ion is removed from its normal site,

V_o anion vacancy,

M_i^{++} interstitial cation.

Other point defects such as charged vacancies and associated vacancies and ions are left out of the discussion. The concentration of defects and ions are correlated by equilibria such as

$$M^{++} \rightleftharpoons M_i^{++} + V_c; \quad V_c + V_o \rightleftharpoons zero .$$

The second reaction describes the formation and annihilation of vacancies. This reaction can only take place at outer surfaces and grain boundaries, but this does not imply that the equilibrium concentrations cannot be attained inside the crystal. When a small temperature increase is imposed on a crystallite which is in equilibrium, vacancies are formed at the outer surface until the new value of the equilibrium constant $K_s = [V_c] \times [V_o]$ is reached. This will cause a diffusional flux of vacancies to the interior of the crystallite until K_s has obtained its new value throughout the crystallite. Similar diffusional fluxes of vacancies occur during sintering, but the origin of the vacancy gradients is quite different from the case above. The vacancy concentration in a surface layer of a solid depends on the local shape of the surface. This is because the surface free energy of small solid particles is a substantial part of the total free energy and cannot be omitted in thermodynamic considerations on small particles. According to Kelvin [1858], the smaller the particles the higher the vapour pressure and by a similar reasoning it can be proved that for small solid particles the vacancy concentration decreases with decreasing particle size. This proof will now be given.

Consider a small cube of a crystalline element containing N atoms of the element and with no vacancies. The Gibbs free energy of this crystallite decreases when vacancies are formed and the equilibrium concentration of vacancies is attained, when the Gibbs free energy has minimum value (at constant T, P and constant number of atoms). The formation of n vacancies causes a volume increase of $n\Omega$ in which Ω is the volume increase of the crystallite when one vacancy is formed. From this we see that the surface increase of the cube will be equal to $(4/a) n\Omega$ in which a is the size of the cube (for a sphere the surface increase would be $(2/\varrho)n\Omega$ and for a spherical pore $-(2/\varrho)n\Omega$). The increase of surface energy due to the formation of the vacancies is consequently $(4/a)n\Omega\gamma$ (the surface energy γ is assumed to be independent of vacancy concentration).

The enthalpy of formation of one vacancy is denoted by h and as long $n \ll N$ the total enthalpy increase is given by nh. The increase of configurational entropy is calculated from the Boltzmann relation $S_{conf} = k \ln W = k \ln \{(N+n)/N!n!\}$. Furthermore, let us assume that the thermal entropy increases per vacancy by an amount Δs_{th}. We may now write for the increase of Gibbs free energy

$$\Delta G = nh + \frac{4}{a} n\Omega\gamma - nT\Delta s_{th} - kT \ln \frac{(N+n)!}{N!n!}. \qquad (6.1)$$

The number of vacancies when equilibrium is reached is found from $(\partial \Delta G/\partial n)_{T,P,N} = 0$:

$$h + \frac{4}{a}\Omega\gamma - T\Delta s_{th} + kT \ln \frac{n}{N+n} = 0. \qquad (6.2)$$

The concentration of vacancies C_v is most adequately expressed by the fraction of sites which are empty $C_v = n/(N+n)$.

From eq. (6.2) we find

$$C_v = \exp\{-H/RT + \Delta S_{th}/R\} \exp\{-(4\gamma/a)V_m/RT\} \quad \text{or}$$

$$C_v = C_v(a \rightarrow \infty) \exp\{-(4\gamma/a)V_m/RT\}. \tag{6.3}$$

H is the enthalpy of formation of N_0 vacancies in a very large crystal such that the number of vacancies is much smaller than the number of atoms. Likewise ΔS_{th} is the increase of thermal entropy by the formation of these N_0 vacancies and V_m is the volume of N_0 vacancies and is approximately equal to the molar volume of the solid (N_0 is Avogadro's number). We have just seen that for a cube (d Surface/d Volume)$=4/a$ and for a pore (d Surface/d Volume)$= -2/\varrho$. The vacancy concentration at the surface of a pore is thus found to be

$$C_v(\varrho) = C_v(\varrho \rightarrow \infty) \exp\{(2\gamma/\varrho)V_m/RT\}. \tag{6.4}$$

The difference between $C_v(\varrho)$ and $C_v(\varrho \rightarrow \infty)$ is about 1% of the value of C_v for a pore with a radius of 10^{-5} cm. This is calculated from expression (6.4) by inserting the following values: $2\gamma \approx 1000$ erg/cm^2, $V_m \approx 10$ cm^3 and $RT \approx 10^{11}$ erg, or $T \approx 1200°$K. There is thus a small difference in vacancy concentration between the material at the surface of the pore and the outer surface of the crystal, resulting in a diffusional stream of vacancies by which the pore is annihilated. During the actual sintering process, hollow surfaces are the vacancy sources and the grain boundaries are the vacancy sinks.

Pores lying within the grains will disappear more rapidly when volume diffusion is enhanced, but in the case of pores lying on grain boundaries this is less evident because they can be annihilated by surface and grain boundary diffusion as well. What is the nature of grain boundary diffusion? If it is the movement of ions in the narrow region where the two adjacent crystals do not fit, it is not likely that the defect structure of the solid will have much influence on this type of material transport. However, the nature of grain boundary diffusion may in fact be volume diffusion in the region of grain boundaries. In that case grain boundary diffusion depends on the defect structure of the solid in a similar way as volume diffusion. The existence of enhanced volume diffusion in the region of the grain boundaries has been established experimentally by Oishi and Kingery [1960] and suggested by Wuensch and Vasilos [1964]. It is my opinion that this enhanced diffusion is due to impurities which are dissolved but absorbed at the grain boundaries like MgO in Al_2O_3. This phenomenon is usually referred to as solute segregation and has been extensively investigated, e.g. by Cahn [1962] and Westbrook [1967]. The higher concentration of MgO in the grain boundary regions of polycrystalline alumina will increase the oxygen vacancy concentration and enhance volume diffusion of oxygen ions in that region.

For the sake of convenience only the problem of vacancies emitted by a spherical pore within a large crystal is treated in this paper and the rate of pore annihilation is calculated as a function of small deviations from stoichiometry. The dependence of volume diffusion on composition so derived is thought to be a more general property of the material and will be applied quantitatively to the more complex sintering of powder compacts where there are a large number of pores of different sizes within the crystals and on the grain boundaries.

2.2. The vacancy flux emitted by a spherical pore in a metal crystal

The outward flux of vacancies Φ emitted from the pore surface can be expressed in terms of the diffusion constant and the concentration of vacancies. According to Fick's first law

$$\Phi = 4\pi r^2 I = -4\pi r^2 D \frac{\partial C(r)}{\partial r} \tag{6.5}$$

where Φ is the total flux of vacancies and r is the distance to the center of the pore.

In solid-state chemistry the concentrations of ions and defects are very often expressed by

$$[Y] = C_y = \frac{\text{number of Y species}}{\text{number of available sites}} .$$

This definition is used in this chapter and consequently the dimension of D in Fick's law is moles $cm^{-1} sec^{-1}$.

The differential equation (6.5) is easy to solve when it is assumed that Φ is independent of r in the steady state. The general solution is $C(r) = A + B/r$. The constants A and B follow from the boundary conditions $C(\varrho) = C\exp\{(2\gamma/\varrho)(\Omega/kT)\} = C[1 + (2\gamma/\varrho)(\Omega/kT)]$ and $C(r \to \infty) = C$. Consequently $C(r)=C[1+(2\gamma/r)\ (\Omega/kT)]$. The flux is found to be $\Omega=4\pi\ (2\gamma\ \Omega/kT)\ DC=4\pi ADC$, where $A=2\gamma\ \Omega/kT\approx 10^{-7}$ cm. This simple solution is only justified when the approximation exp $\{(2\gamma/\varrho)\ (\Omega/kT)\}=1+2(\gamma/\varrho)\ (\Omega/kT)$ is valid. As mentioned before, the value of $(2\gamma/\varrho)(\Omega/kT)$ is 10^{-2} at the sintering temperature for pores with a radius of 10^{-5} cm. For extremely small pores, the vacancy flux becomes slightly dependent on ϱ and the vacancy concentration at a distance r from the pore increases slightly with time. The material transport involved in the annihilation of these extremely small pores is relatively small and therefore there is no need to modify the solution for the very last part of the process.

2.3. The vacancy flux emitted by a pore in an ionic solid MO

The case of ionic solids is far more complicated than that of metals because there are at least two different ion types which must be transported during sintering. The simplest case one can think of is an ionic solid MO in which we have only cation vacancies V_c and anion vacancies V_o. As a result of the dissolution of small amounts of oxides which have a cation-to-anion ratio differing from that of the host lattice extra vacancies are introduced in the lattice. Dissolution of small amounts of mono-valent oxide L_2O introduces extra oxygen vacancies or interstitial cations while a trivalent oxide A_2O_3 gives extra cation vacancies. If MO is an oxide in which the cations are considerably smaller than the oxygen ions, then it is not likely that dissolution of A_2O_3 will give rise to the formation of interstitial oxygen ions. The number of vacancies can thus drastically be increased by dissolution of impurity oxides but the relation $K_s=[V_c]\times[V_o]=C_c\times C_o$ still holds for the solid solution. At the surface of a pore the value of K has a somewhat increased value and obeys the Kelvin equation $K(\varrho)=K\exp(2\gamma/\varrho)\times(\Omega/kT)$ where Ω is now the volume of a vacancy pair V_c+V_o. Both cation and anion vacancies 'are emitted by the pore and the two fluxes Φ_c and Φ_o must correspond with the cation-to-anion ratio of the material, because the pore is filled up with electrically neutral material and equal amounts of cation and anion sites. If one of the ion species tends to diffuse more quickly this gives rise to an electric field $E(r)$ which opposes the flux of the fast moving species and promotes the flux of the slow moving species such that in the steady state the ratio of Φ_c and Φ_o has its proper value. If the composition of the solid solution is $(1-x)$ MO$+x$L$_2$O it follows that $\Phi_c/\Phi_o=1+x$. Taking into account the presence of an electric field the fluxes are expressed by

$$\Phi_c = (1+x)\Phi_o = 4\pi r^2 I_c = 4\pi r^2 \left[-D_c \frac{\partial C_c(r)}{\partial r} - \mu_c C_c(r) E(r) \right] \quad (6.6)$$

$$\Phi_o = 4\pi r^2 I_o = 4\pi r^2 \left[-D_o \frac{\partial C_o(r)}{\partial r} + \mu_o C_o(r) E(r) \right]. \quad (6.7)$$

The mobilities μ_c and μ_o are correlated with the diffusion constants D_c and D_o by the Nernst-Einstein equation $\mu kT = |q| D$ where $|q|$ is the charge of the moving species. Eliminating $E(r)$ from expressions (6.6) and (6.7) after having substituted $\mu = |q| D/kT$ one obtains:

$$\Phi_c D_o C_o(r) + \Phi_o D_c C_c(r) = 4\pi r^2 \left[-D_c \frac{\partial C_c(r)}{\partial r} D_o C_o(r) - \right.$$

$$\left. D_o \frac{\partial C_o(r)}{\partial r} D_c C_c(r) \right]. \quad (6.8)$$

Just as in the case of metals Φ_c and Φ_o are assumed to be independent of r, which means that $C_o(r)$ and $C_c(r)$ are independent of time. By analogy with the solution for a pore in a metal we choose the following solutions for $C_o(r)$ and $C_c(r)$

$$C_o(r) = C_o\left(1 + \frac{a}{r}\right) \quad \text{and} \quad C_c(r) = C_c\left(1 + \frac{b}{r}\right) \quad (6.9)$$

with $A = a + b = 2\gamma(\Omega/kT)$.

Equation (6.8) is then transformed to

$$\Phi_c D_o C_o \left(1 + \frac{a}{r}\right) + \Phi_o D_c C_c\left(1 + \frac{b}{r}\right) =$$

$$4\pi \left[D_c C_c D_o C_o \left(1 + \frac{a}{r}\right) b + D_o C_o D_c C_c \left(1 + \frac{b}{r}\right) a \right]. \quad (6.10)$$

As a/r and $b/r \ll 1$, eq. (6.10) is reduced to

$$\Phi_c D_o C_o + \Phi_o D_c C_c = 4\pi A D_o D_c C_o C_c. \quad (6.11)$$

Together with $\Phi_c = (1+x)\Phi_o$ we find the following expression for the oxygen vacancy flux

$$\Phi_o = \frac{4\pi A D_o D_c C_o C_c}{(1+x)D_o C_o + D_c C_c}. \quad (6.12)$$

References p. 238

The ratio of a and b depends on the distribution of L^+ ions. The electric field $E(r)$ is calculated from eqs. (6.6), (6.7) and (6.9):

$$E(r) = \frac{kT}{r^2|q|} \frac{D_c C_c b - (1+x)D_o C_o a}{D_c C_c + D_o C_o (1+x)}. \tag{6.13}$$

The concentration of L^+ ions, C_L, and the vacancy concentrations are correlated by

$$C_o(r) = \tfrac{1}{2}C_L(r) + C_c(r). \tag{6.14}$$

Together with eq. (6.9) the dependence of C_L on r is found to be

$$C_L(r) = C_L \left(1 + \frac{C_o a - C_c b}{2C_L r}\right). \tag{6.15}$$

By inspecting eqs. (6.13) and (6.15) it is seen that it is impossible for $E(r)$ $=0$ and $C_L(r)$ to be independent of r at the same time, (unless $D_c = D_o$ and $x=0$). This was the starting point for the calculations of Johnson [1967] which led to erroneous results.

We now return to equation (6.12) which is simplified for small values of x to

$$\Phi = \frac{4\pi A D_o D_c C_o C_c}{D_o C_o + D_c C_c} = \frac{4\pi A}{(1/D_o C_o) + (1/D_c C_c)}. \tag{6.16}$$

For a stoichiometric material, where $C_o = C_c = C$, eq. (6.16) is further simplified to $\Phi = 4\pi A D C$ in which $1/D = 1/D_o + 1/D_c$. This is the same expression as the one obtained for metals. According to eq. (6.16) the maximum flux is obtained when $D_o C_o + D_c C_c$ has the minimum value, that is when $D_o + D_c \, \partial C_c/\partial C_o = 0$. As $C_c \times C_o = K_s$ the maximum value is obtained when $D_o C_o = D_c C_c$ and $\Phi_{max} = 2\pi A D_o C_o = 2\pi A D_c C_c$.

For a graphical representation of eq. (6.16) we put it first on a relative scale:

$$\frac{\Phi}{\Phi_{stoich}} = \frac{(D_o + D_c) K_s^{\frac{1}{2}}}{D_o C_o + D_c C_c}.$$

When $D_c C_c \gg D_o C_o$; $\log \dfrac{\Phi}{\Phi_{stoich}} = \log \left(1 + \dfrac{D_o}{D_c}\right) K_s^{\frac{1}{2}} - \log C_c.$

When $D_o C_o \gg D_c C_c$; $\log \dfrac{\Phi}{\Phi_{stoich}} = \log \left(1 + \dfrac{D_c}{D_o}\right) K_s^{\frac{1}{2}} - \log C_o.$

When $D_o C_o = D_c C_c$; $\log \dfrac{\Phi}{\Phi_{\text{stoich}}} = \log\left(1 + \dfrac{D_o}{D_c}\right) K_s^{\frac{1}{2}} - \log C_c - \log 2$

$$= \log\left(1 + \dfrac{D_c}{D_o}\right) K_s^{\frac{1}{2}} - \log C_o - \log 2.$$

In Fig. 6.1 $\log(\Phi/\Phi_{\text{stoich}})$ is plotted versus $\log C_c$ and $\log C_o$. The plot goes asymptotically to the straight lines which represent the correlation between $\log(\Phi/\Phi_{\text{stoich}})$ and $\log C_c$, $\log C_o$ for large deviations from stoichiometry.

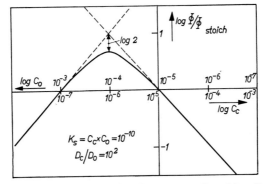

Fig. 6.1. Plot of $\log(\Phi/\Phi_{\text{stoich}})$ versus $\log C_c$ and $\log C_o$.

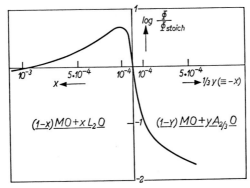

Fig. 6.2. Plot of $\Phi/\Phi_{\text{stoich}}$ versus the composition parameters x and y.

A plot of $\log(\Phi/\Phi_{\text{stoich}})$ versus x, the amount of impurity oxide added, is more useful for the daily practice of sintering (Fig. 6.2). It is seen from this plot that whenever $D_c > D_o$ sintering is promoted by making the material anion-deficient. After the maximum has been passed the sintering rate is slowly decreased when more impurity oxide is added. On the other hand the sintering rate is drastically decreased when the material is made cation-deficient.

The tendency to form interstitial cations will now be considered and its influence on the vacancy flux will be derived. The number of interstitial cations will be much larger in the anion-deficient region. This is easily seen from the equilibrium reaction $M^{++} \rightleftarrows M_i^{++} + V_c$ with $k_i \cdot [M^{++}] = [M_i^{++}] \times [V_c]$ or $k_i \cdot C_M = C_i \cdot C_c$.

We assume that the molar volume of the material and the surface free energy are not changed by the formation of interstitial cations. In that case k_i does not depend on pore size. We further assume that the reaction $M^{++} \rightleftarrows M_i^{++} + V_c$ can take place in the bulk of the material. Therefore $k_i(\partial C_M/\partial r) = C_c(\partial C_i/\partial r) + C_i(\partial C_c/\partial r)$. Because $k_i \ll C_c$ or C_i and $\partial C_M/\partial r$, $\partial C_i/\partial r$, $\partial C_c/\partial r$ are of comparable magnitude, we find $C_c(\partial C_i/\partial r) + C_i(\partial C_c/\partial r) = 0$. From this result we derive $\Phi_c/\Phi_i = -D_c C_c/D_i C_i$. Together with $(\Phi_c - \Phi_i)/\Phi_0 = 1 + x$ and eq. (6.11) $\Phi_c D_0 C_0 + \Phi_o D_c C_c = 4\pi A D_0 C_0 C_c$ we find for small values of x

$$\Phi = \frac{4\pi A}{1/D_o C_o + 1/(D_c C_c + D_i C_i)}.$$

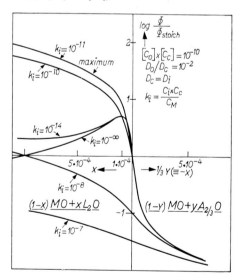

Fig. 6.3. Plot of $\log(\Phi/\Phi_{\text{stoich}})$ versus the composition parameters x and y for different values of the equilibrium constant k; of the Frenkel equilibrium $M^{++} \rightleftarrows M_i^{++} + V_c$.

The influence of interstitial cations on the flux can be calculated from this expression. An example is given in Fig. 6.3. A slight tendency to form interstitial cations will increase the flux in the anion-deficient region where $D_o C_o > D_c C_c$. The reason is that there are two mechanisms by which the cations can diffuse, a vacancy and an interstitial mechanism. The optimal composition for fast sintering is reached when $D_o C_o \approx D_c C_c + D_i C_i$.

3. Experiments and results

As so many other factors are involved in the sintering process, it is rather difficult to correlate experimentally the rate of densification of a powder compact during sintering uniquely to small deviations from stoichiometry. For example the morphology of the powder compact has a particularly strong influence. Generally speaking the sintering rate increases with increasing specific surface. Small amounts of a dispersed second phase by which grain growth is inhibited decrease the sintering rate considerably.

The most appropriate method to study experimentally the correlation between sintering and composition will now briefly be described. Oxides in which the cations are considerably smaller than the oxygen ions are sintered best when they are anion-deficient. Therefore quite a large amount of a slightly anion-deficient material is made by a classical ceramic technique that is mixing the raw materials in a ball mill, prefiring and ball-milling again. A wet-chemical method of preparation is more elegant e.g. spray-drying of a solution of soluble salts like sulphates. In this way dried sulphates are obtained in which the cations are distributed on an atomic scale. Upon heating, the sulphates decompose into a powder of very homogeneous composition. The powder is prefired at a fairly high temperature in order that powders are obtained with a particle size of a few tenths of a micron. Small portions of this anion-deficient powder are now mixed in an agate mortar with an impurity oxide in a reactive form, which upon dissolution introduces cation vacancies in the host lattice. By prolonged heating at a temperature slightly below the prefiring temperature, the impurity oxide is dissolved while the original powder morphology is preserved. The fact that dissolution of the impurity oxide introduces cation vacancies in the bulk material favours fast dissolution and homogenization of the solid solution. If these precautions are not taken and the solid solutions are prepared by mixing the original reactive raw materials and the impurity oxides together, then the anion-deficient powder is more coarsened after prefiring and consequently has a

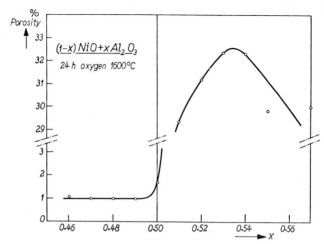

Fig. 6.4. Porosities of sintered products of composition $(1-x)$ NiO $+$ xAl₂O₃.

larger particle size than the cation-deficient material. This is undesirable because the sintering rate depends on the particle size of the powder compact. In most of our experiments only final densities of the samples were measured, but sintering temperatures as well as sintering times were varied. Some of the results will briefly be discussed. Other examples have been published previously: Reijnen [1968a, 1969].

a) $NiAl_2O_4$+excess NiO   (anion-deficient)
      $NiAl_2O_4$+excess Al_2O_3 (cation-deficient).

The difference in sintering behaviour is so pronounced in this case that whatever experimental procedure is chosen, a difference will be observed both in the sintering rate and final density. One set of sintering densities is represented in Fig. 6.4. The anion-deficient samples have porosities of $\sim 1\%$ and the cation deficient samples of more than 30%.

b)   NiO+small amounts of Li_2O (anion-deficient)
      NiO+small amounts of Al_2O_3 or Fe_2O_3 (cation-deficient).

This system is more complicated than the preceding one as the dissolution of impurity oxides is partly compensated by a change of valancy of the cations rather than by the formation of vacancies.

The dissolution of Li_2O is compensated to the extent of 90% by the formation of Ni^{+++} and only 10% of the Li_2O dissolved is effective with

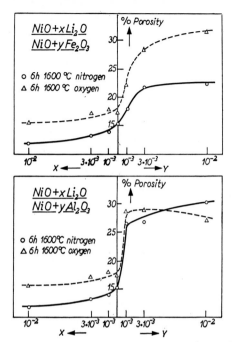

Fig. 6.5. Porosities of sintered products of composition NiO + x Li₂O, NiO + yFe₂O₃, NiO + yAl₂O₃.

regard to the formation of anion vacancies (see Bosman and Crevecoeur [1966]).

When the oxygen partial pressure is decreased the anion vacancy concentration increases according to

$$\text{Ni}^{+++} + \tfrac{1}{2}\text{O}^{--} \rightleftarrows \text{Ni}^{++} + \tfrac{1}{4}\text{O}_2 + \tfrac{1}{2}\text{V}_\text{o}.$$

Samples containing small amounts of Al_2O_3 or Fe_2O_3 have considerably higher porosities than samples containing Li_2O because they are cation-deficient, but the addition of Fe_2O_3 is less effective than the addition of Al_2O_3. This again is explained by a change of valency of part of the ferric ions and a corresponding decrease of the number of cation vacancies according to

$$\text{Fe}^{+++} + \tfrac{1}{2}\text{O}^{--} + \tfrac{1}{4}\text{V}_\text{c} \rightleftarrows \text{Fe}^{++} + \tfrac{1}{4}\text{O}_2 .$$

The sintering results are in agreement with this view (Fig. 6.5). The sintering of NiO doped with Al_2O_3 does not depend on the oxygen partial pressure,

References p. 238

but NiO doped with Fe_2O_3 is sintered to a higher density in nitrogen than it is in oxygen. Some knowledge of the solid-gas equilibrium might thus be of help in the study of the sintering phenomena of a system.[1]

c) Y_2O_3 + small amounts of CaO   (anion-deficient)
    Y_2O_3 + small amounts of ThO_2 (cation-deficient).

It is not a priori evident that the oxygen vacancies are the slower diffusing species in this system. The Y^{3+} ion has a radius of 0.92 Å and the oxygen ion of 1.32 Å but the Y^{3+} ion has a higher charge. From sintering experiments, however, it appears that the samples containing CaO sinter faster than those containing ThO_2, implying $D_c > D_o$. By prolonged heating, however, the ThO_2-doped yttria is sintered practically to full density (Fig. 6.6). Jorgensen and Westbrook [1967] succeeded in sintering yttria to full density.

The materials sintered by Jorgensen contained a large amount of dissolved ThO_2 up to 10 mole %. The sintering behaviour of cation-deficient yttria will be dealt with in the discussion presented in the next section.

4. Discussion

Volume diffusion during sintering of ionic solids proves to be strongly dependent on the defect structure of the solid and it has been found that volume diffusion and sintering is enhanced when the concentration of the slowest moving defects is increased. Initial sintering of powders with a very small particle size, say < 1000 Å, might take place predominantly by surface diffusion, in which case the dependence of initial sintering rate on the deviation from stoichiometry could be quite different. This point remains to be investigated. Together with a decreased sintering rate of cation-deficient powders it is observed in many cases that grain growth too is decreased. For example $NiAl_2O_4 + Al_2O_3$ shows no tendency to sinter at 1600°C but the grains remain in the submicron range, whereas $NiAl_2O_4 + NiO$ at 1600°C sinters very well and has grains of approximately 5µ after 24 h (Reijnen [1968a]). A similar correlation between sintering rate and grain growth is found when a dispersed inert second phase is added to the material. Grain growth is inhibited by this measure but sintering is almost stopped. As the

---

[1] Especially the equilibria between ferrite systems and oxygen have been investigated. See e.g. Paladino [1959, 1960], O'Bryan [1966], Schmalzried [1966] and Reijnen [1965, 1968].

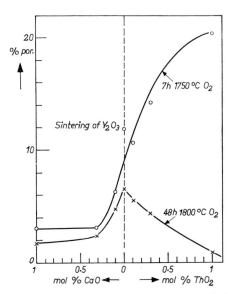

Fig. 6.6. Porosities of sintered products of composition $Y_2O_3 + xCaO$ and $Y_2O_3 + yThO_2$.

grain boundaries are the sinks for vacancies during the sintering process, it follows that the slowly sintering material with the small grains has not yet lost its sintering power. Sintering and grain growth proceed slowly but steadily and it only takes more time to sinter the material to full density. For example, finely dispersed ThO_2 (0.1 % or less) and especially finely dispersed $BaSO_4$ in small amounts (0.01 % or less) decrease the sintering rate and grain growth of ferrites, but after prolonged heating the density and micro-structure are practically equal to those of the material sintered without a second inert phase (Reijnen, unpublished results). The only difference in sintering behaviour is that the material with no second phase is sintered to its final density in a much shorter time.

The phenomenon of decreased sintering rate due to empediment of grain growth is demonstrated in Fig. 6.6 where it is seen that ThO_2-doped yttria after being sintered for a long time has almost full density. In this particular case grain growth is inhibited by solid segregation at the grain boundaries of the dissolved ThO_2 according to Jorgensen and Westbrook, but in my opinion the unfavourable defect structure might as well be the reason for the decreased sintering rate and decreased grain growth (see pp. 235 and 236). Sometimes this behaviour is disturbed because discon-

References p. 238

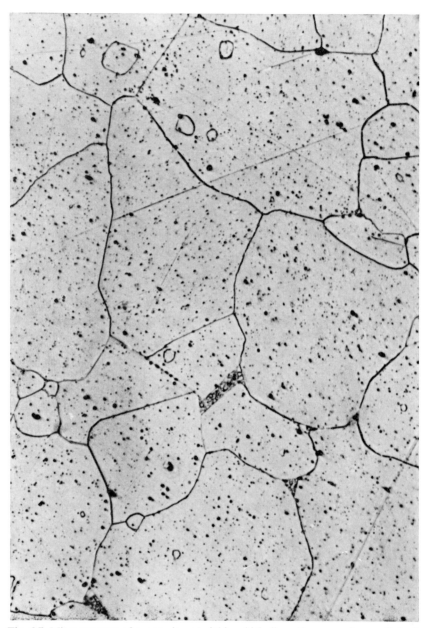

Fig. 6.7. Microstructure of a ceramic material in which discontinuous grain growth took place. The microstructure is characterized by large crystals with erratic grain boundaries and many pores within the crystals. Magnification 230 ×. The material is a NiZn ferrite and discontinuous grain growth is provoked by 0.05 weight % of dispersed $BaSO_4$.

tinuous grain growth takes place. When grain growth is very strongly inhibited, as in $NiAl_2O_4$ + excess Al_2O_3 or in ferrites containing $\sim 0.1\%$ dispersed $BaSO_4$, it may happen that somewhat larger crystallites are nucleated in the matrix of extremely fine grains. The driving force for the growth of these crystallites is extremely large because the driving force for grain growth of a crystal is inversely proportional to the linear size of the neighbouring small crystallites. In a short time, most grain boundaries then disappear and the material loses its sintering power. The large grains are erratically shaped and many pores are captured within the grains (Fig. 6.7). Two questions or problems remain to be answered with respect to the foregoing discussion.

1. Why is sintering decreased when grain growth is inhibited with a dispersed inert second phase?

2. Why is grain growth decreased when sintering is impeded by an unfavourable defect structure?

The first problem is difficult to solve on an experimental basis. If the grain boundaries are pinned by a second phase, the vacancies must be precipitated homogeneously and simultaneously on *all* grain boundaries, as otherwise no shrinkage can occur. It takes time to distribute the vacancies homogeneously over the grain boundaries by volume or grain boundary diffusion. The sintering rate will be much higher if the vacancies supplied to a grain boundary can directly be digested. This is possible if the grains can be slightly displaced with respect to each other (grain-boundary sliding). In this case precipitation of vacancies must be homogeneous at each particular grain boundary but the rate of precipitation can be different for different grain boundaries. It is now put forward as a hypothesis that grain-boundary sliding is essential for fast sintering as normally observed and also, that, due to the presence of a dispersed second phase, grain-boundary sliding is suppressed for purely geometrical reasons. This explains the influence of an inert second solid phase on sintering.

The second question to be answered is why grain growth is decreased when sintering is impeded by an unfavourable defect structure. This is due to pores which inhibit grain growth. Pores lying on a grain boundary act just like a second phase, but pores are less effective as they can be swept by the grain boundaries (see e.g. Speight and Greenwood [1964]). In Fig. 6.8 it is shown that the shape of a pore changes when it exerts a pull on a grain boundary. Pores and grain boundaries of this shape are often seen in ceramic microstructures (Reijnen [1968a]). As the pore has two different radii of

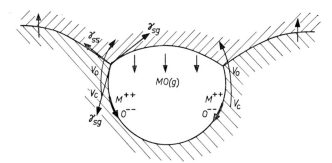

Fig. 6.8. Shape of a pore on a moving grain boundary. Material transport from one side of the pore to the other can take place by a) volume diffusion of vacancies, b) surface diffusion of ions and c) vapour phase transport.

curvature the vacancy concentrations at the two pore surfaces are also different, resulting in a diffusional flux of vacancies. The pore will move along with the grain boundaries but in the meantime exerts a drag on it. The smaller the pore, the larger the gradient between the two pore surfaces and the easier will the pore be swept by the grain boundary. When volume diffusion is suppressed by an unfavourable defect structure, the mobility of the pores is slowed down, and, as the grain boundaries are pinned to the pores, grain growth is inhibited. Thus, because both grain growth and sintering are determined by the same diffusion process, the results after prolonged sintering will be identical irrespective of the defect structure. The exceptions to this rule are caused by discontinuous grain growth or by the fact that material transport from one side of the pore does not take place by volume diffusion but by surface or vapour phase transport. An example of such a case is found in the sintering behaviour of cation-deficient ferrites sintered in oxygen. These materials have peculiar microstructures (Fig. 6.9). The pores are found exclusively on the grain boundaries and the grains are rather large. The explanation for this phenomenon has previously been published (Reijnen [1968a]), and will briefly be repeated here. The ferrite material is in equilibrium with the oxygen atmosphere:

$$Fe^{+++} + \tfrac{3}{8}V_c + \tfrac{1}{2}O^{--} \rightleftarrows Fe^{++} + \tfrac{1}{4}O_2'$$

and the value of the equilibrium constant increases with increasing pore size in accordance with the Kelvin equation discussed in Section 2.1. Material transport from one side of the pore to the other in Fig. 6.10, is made possible in this case by transport of oxygen via the gas phase and volume diffusion of cations. This volume diffusion of cations is rather rapid as the material is

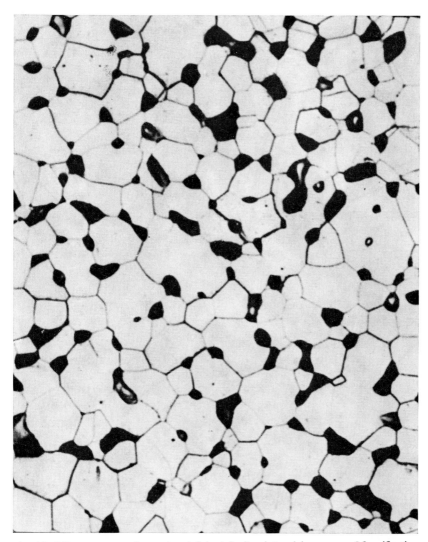

Fig. 6.9. Microstructure of a cation-deficient ferrite sintered in oxygen. Magnification 900 ×. The microstructure is characterized by pore-free grains and large pores on the grain boundaries. Pores are swept by the grain boundaries in this material by enhanced material transport from one side of the pore to the other. The mechanism of this transport is shown in Fig. 6.10.

References p. 238

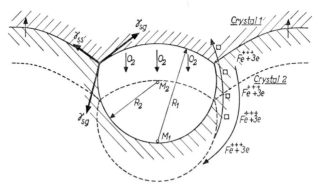

Fig. 6.10. The mechanism of material transport from one side of the pore to the other in cation-deficient ferrites sintered in an oxidizing atmosphere.

cation-deficient. The pores are now easily swept by the grain boundaries resulting in pore free and large grains. This proves that grain growth in a single-phase system is governed by the mobility of the pores and less by the defect structure of the solid.

References

BOSMAN, A. J. and C. CREVECOEUR, 1966, Phys. Rev. **144**, 763.

CAHN, J. W., 1962, Acta Met. 789.

JOHNSON, D. L., 1967, Sintering, in: G. C. Kuczynski, N. A. Hooton and C. F. Gibbon, eds. (Gordon and Breach, New York) pp. 393–400.

JONES, J. T., P. K. MAITRA and I. B. CUTLER, 1958, J. Am. Ceram. Soc. **41**, 353.

JORGENSEN, P. J., 1965, J. Am. Ceram. Soc. **48**, 207.

JORGENSEN, P. J. and J. H. WESTBROOK, 1967, J. Am. Ceram. Soc. **50**, 553.

MARSHALL Jr., P. A., D. P. ENRIGHT and W. A. WEYL, 1954, Proc. Intern. Symp. on the Reactivity of Solids (Gothenburg) pp. 273–284.

O'BRYAN, H. M., F. R. MONTFORTE and R. BLAIR, 1965, J. Am. Ceram. Soc. **48**, 577.

OISHI, Y. H. and W. D. KINGERY, 1960, J. Chem. Phys. **33**, 480.

PALADINO, A. E., 1960, J. Am. Ceram. Soc. **43**, 183.

READEY, D. W., 1966, J. Am. Ceram. Soc. **49**, 366.

REIJNEN, P. J. L., 1968a, Science of Ceramics **4**, 169.

REIJNEN, P. J. L., 1968b, Philips Res. Rept. **23**, 151.

REIJNEN, P. J. L., 1969, Proc. 6th Intern. Symp. on the Reactivity of Solids (Schenectady, 1968) p. 99.

SCHMALZRIED, H. and J. D. TRETJAKOW, 1966, Ber. Bunsenges. Physik Chem. **70**, 180.

SPEIGHT, M. V. and G. W. GREENWOOD, 1964, Phil. Mag. **6**, 683.

THOMSON, W. (Lord KELVIN), 1858, Proc. Roy. Soc. (London) **9**, 255.

WESTBROOK, J. H., 1967, Science of Ceramics **3**, 263.

WUENSCH, B. J. and T. VASILOS, 1966, J. Am. Ceram. Soc. **49**, 433.

CHAPTER 7

MÖSSBAUER SPECTROSCOPY
OF NONSTOICHIOMETRIC PHASES

N. N. GREENWOOD

Department of Inorganic Chemistry, University of Newcastle upon Tyne,
Newcastle upon Tyne, England

1. Introduction: The Mössbauer effect

Mössbauer spectroscopy, or nuclear resonance fluorescence spectroscopy, depends on the recoilless emission of γ-rays and their resonant reabsorption, an effect discovered accidentally by Mössbauer [1958]. The technique has developed rapidly during the last decade and has made numerous and diverse contributions in several areas of physics and chemistry. This chapter explores the ways in which Mössbauer spectroscopy can be used in the study of nonstoichiometric phases and indicates some of the results which have been obtained. The examples chosen are illustrative rather than comprehensive and it is not intended to review all the results which have so far been obtained. It will become apparent that Mössbauer spectroscopy is a powerful and versatile new technique which gives much valuable information on solid state systems; it will also emerge that the technique has inherent limitations which restrict its applicability in certain cases.

The physical basis of the Mössbauer effect has recently been reviewed in detail by Goldanskii and Herber [1968] and by Greenwood [1970] and it is therefore only necessary to indicate briefly the underlying principles and terminology.

In order to observe recoilless emission of γ-rays four conditions are necessary:
(i) The γ-ray should have comparatively low energy ($E_\gamma \sim 10$–150 keV) since the probability of recoil-free emission diminishes as $\exp(-E_\gamma{}^2)$;

References p. 267

(ii) The decaying nuclide should be firmly bound in a solid so that the recoil energy, E_R, is a property of the matrix rather than of the individual atoms and is thus negligible: $E_R = E_\gamma^2/(2Mc^2)$;

(iii) The γ-transition should have an appropriate half-life (in the range 1 μsec–1 psec) in order to obtain lines of observable Heisenberg line-width Γ:
Γ (eV) $= 4.562 \times 10^{-16}/t_{\frac{1}{2}}(\text{sec})$;

(iv) There should be a low internal conversion coefficient to ensure that the γ-transition manifests itself as a γ-photon and not as a conversion electron.

These conditions are fulfilled by the excited states of about 70 isotopes of 40 different elements in the Periodic Table, but only about eight of these have so far been used to study nonstoichiometric phases. The most fully investigated isotope is ^{57}Fe, but defect studies have also been published using ^{119}Sn and there are isolated papers on ^{121}Sb, ^{125}Te, ^{151}Eu, ^{166}Er, ^{170}Yb and ^{197}Au. Undoubtedly, the use of these and other isotopes will increase in the future.

The significance of recoilless emission of γ-rays is that it provides energy quanta of unparalleled precision, thus enabling the resolution of hyperfine interactions which were previously obscured, and the detection of minute changes in energy which were formerly unobservable. In the typical case of ^{57}Fe ($E_\gamma = 14.41$ keV, $t_{\frac{1}{2}} = 97.7$ nsec, $\Gamma = 4.67$ neV) the ratio of the experimental line width (2Γ) to the total γ-ray energy is 0.648×10^{-12}. If such a γ-

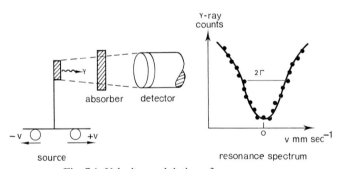

Fig. 7.1. Velocity modulation of γ-ray energy.

photon emerging from the source strikes an identical atom in an absorber, it will be resonantly reabsorbed; but if the ^{57}Fe atoms in the absorber are not in an identical physical and chemical environment to those in the source, the energy of the γ-transition will not exactly match and no absorption will occur.

To restore resonance, energy must be supplied and this is done by moving the source towards the absorber. A speed of $0.648 \times 10^{-12}c$ (i.e. 0.194 mm/ sec) will shift the resonance position by one full linewidth at half height. A Mössbauer spectrometer imparts a series of Doppler velocities in sequence to the source and at some particular velocity maximum resonant absorption occurs. This is schematically illustrated in Fig. 7.1. It is customary to use electromechanical devices and servo-mechanisms to impart the Doppler velocities and γ-ray counts are recorded in a multichannel analyser, each channel of which corresponds to a small velocity increment.

2. Lattice defects and Mössbauer spectra

Lattice defects, whatever their origin, may alter several of the properties of individual ions and they may also alter the collective properties of the crystal as a whole. Both the intimate, individual changes and the bulk changes will affect various aspects of the Mössbauer spectrum and it is necessary to define what these changes might be. In any spectroscopic technique there are in principle four observables: (a) the number of lines, (b) their position, (c) their shape, and (d) their intensity. In Mössbauer spectroscopy these four observables are dependent on and quantified by five parameters:

δ chemical isomer shift,
Δ electrical quadrupole splitting,
H nuclear hyperfine magnetic field,
Γ line-width at half-height,
f recoil-free fraction.

Each of these parameters have a temperature dependence and a pressure dependence so that a considerable amount of information can be abstracted from the spectrum.

2.1. Chemical isomer shift

The chemical isomer shift, δ, as its name implies, defines the position of the resonance line on the energy scale relative to some arbitrary standard (e.g. hydrated sodium nitroprusside for ^{57}Fe). If more than one chemically distinct atom of a given Mössbauer nuclide is present, then more than one line may be observed. The chemical isomer shift depends, amongst other things, on the difference in s electron density at the resonant nucleus between the compound and the arbitrary standard:

$$\delta = \text{const.} \times \{ |\psi_s(0)|_{std}^2 - |\psi_s(0)|^2 \} .$$

In simple cases it depends at least in part on the oxidation state of the atom. For example, for high-spin iron compounds such as are frequently encountered in solid-state systems, the chemical isomer shift immediately identifies the charge state of the iron. Typical values of δ (relative to $Na_2[Fe(CN)_5NO] \cdot 2H_2O$ taken as zero) are:

Charge state:	Fe^+	Fe^{2+}	Fe^{3+}	Fe^{4+}
δ mm/sec:	~2.3	~1.6	~0.7	~0.3 .

This dependence of chemical isomer shift on charge state finds immediate application in determining ionic valencies in defect solid phases. For example, using electron-capture in ^{57}Co as the precursor of the 14.41 keV Mössbauer level in ^{57}Fe, Mullen [1963] showed that when $^{57}CoCl_2$ was doped into sodium chloride at low concentration, the predominant iron species at high temperatures was Fe^+ substitutionally replacing Na^+. Further, it proved possible to study the dynamics of this species in equilibrium with an Fe^{2+} ion and an associated cation vacancy. From the temperature dependence of the concentration of these two species the energy of interaction of Fe^{2+} and the cation vacancies was found to be 12.7 kcal/mole (1 kcal = 4.184 kJ).

A second illustration of the detection of unusual oxidation states in nonstoichiometric phases is given by the work of Gallagher et al. [1964] who showed that when the compound $SrFeO_{2.5}$ (brownmillerite structure) was oxidized to the perowskite phase $SrFeO_3$ the iron was progressively oxidized from Fe^{3+} to Fe^{4+}.

The elegant work of Hannaford et al. [1965] in detecting Sn^{II} defects in neutron-radiation damaged $Mg_2Sn^{IV}O_4$ is a further example of the use of chemical isomer shifts in identifying altervalent ions in defect solids and studying their removal by thermal annealing in air above 600°.

2.2. Electrical quadrupole interactions

If the Mössbauer nucleus has a nuclear spin quantum number $I > \frac{1}{2}$ in at least one of the two states between which the γ-transition occurs, and if, in addition the nucleus is in a site of symmetry lower than cubic, then the electric field gradient, q, interacts with the nuclear quadrupole moment, Q, to

lift the energy degeneracy of the state and more than one Mössbauer resonance line is observed. For ^{57}Fe and ^{119}Sn, as shown in Fig. 7.2, the ground
state has $I = \frac{1}{2}$ and is unsplit, whereas the excited state has $I = \frac{3}{2}$ and splits into
two levels of separation Δ. When the field is axially symmetrical

$$\Delta = e^2 qQ(1 - \gamma_\infty)$$

where $(1 - \gamma_\infty)$ is the Sternheimer shielding factor. The electric field gradient
can arise from a number of sources of which the most common are (a) the

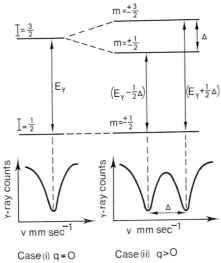

Fig. 7.2. Electric quadrupole interaction.

presence of unpaired d electrons on the atom, (b) the distortion of occupied
lattice site symmetries below ideal cubic, (c) the presence of vacant lattice
sites in the neighbourhood of the Mössbauer nucleus, (d) the presence of
altervalent ions in the lattice. Perhaps the classic case is nonstoichiometric
ferrous oxide. An Fe^{2+} ion in a rigorously octahedral environment shows no
quadrupole splitting because there is no electric field gradient. However, in
the defect wuestite phase $Fe_{1-x}O$ there are x cation vacancies and $(1 - 3x)$
Fe^{3+} ions in addition to the remaining $(1 - 2x)$ Fe^{2+} ions. The majority of
Fe^{2+} ions are therefore in a non-cubic environment and a quadrupole split
spectrum is seen. Early work by Shirane et al. [1962] on quenched samples
was interpreted in terms of a ferrous doublet ($\delta = 1.3$ mm/sec, $\Delta = 0.65$ mm/

sec) and a smaller, unresolved ferric singlet at $\delta \sim 0.5$ mm/sec which gave an asymmetry to the low-velocity quadrupole component. More recent work by Elias and Linnett [1969] has suggested the possibility of electron hopping from Fe^{2+} to Fe^{3+} leading to the observation of averaged chemical isomer shifts and quadrupole splittings.

2.3. Nuclear hyperfine magnetic interactions

If the Mössbauer nucleus is in a magnetic field, Zeeman splitting is observed and this removes the remaining degeneracy of the nuclear energy levels. This is illustrated in Fig. 7.3 for the case of ^{57}Fe. Each level splits into

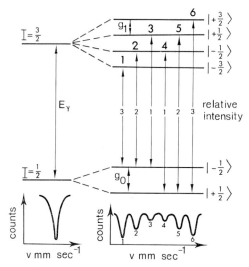

Fig. 7.3. Magnetic hyperfine interactions.

$(2I+1)$ sub-levels and transitions between the ground-state sub-levels and the excited-state sub-levels occur, subject to the selection rules $\Delta m = 0, \pm 1$.

From Fig. 7.3 it is clear that the separation between lines 3 and 4 is $(g_0 - g_1)\beta_N H$ and between all other adjacent pairs of lines is $g_1 \beta_N H$ where g_0 and g_1 are the ground-state and first excited-state nuclear Landé splitting factors, β_N is the Bohr nuclear magneton, $eh/(2Mc)$, and H is the hyperfine magnetic field at the Mössbauer nucleus. The extent of the splitting therefore gives an immediate measure of the magnitude of the hyperfine field and, if there is more than one magnetic site in the unit cell, several superimposed hyperfine spectra may be obtained. An excellent example is afforded by the work of Stearns [1963] on a series of magnetic iron-silicon alloys; the hyperfine magnetic fields at the numerous coexisting iron environments were studied as a function of concentration. The technique is clearly of great use in

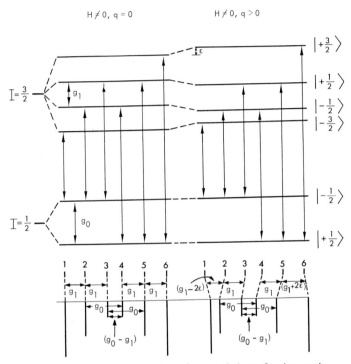

Fig. 7.4. Quadrupole perturbation of magnetic hyperfine interactions.

studying ferromagnetic, antiferromagnetic, and ferrimagnetic exchange interactions and their dependence on temperature.

When both magnetic and quadrupole interactions are present simultaneously, the regular spacing of the lines is perturbed as indicated in Fig. 7.4. In cases where the electric field gradient tensor has axial symmetry this immediately gives the direction of the principal component of the e.f.g. with respect to the magnetic direction. Thus, in Fig. 7.4 the field gradient is positive with respect to the magnetic direction; if it were negative the relative

References p. 267

spacings between lines 1 and 2 on the one hand and lines 5 and 6 on the other would be reversed. It can be seen that the Mössbauer technique provides a most elegant and simple method for studying this alignment of magnetic and electric axes and of studying spin-flop phenomena in the neighbourhood of the Morin temperature. For example, Imbert and Gerard [1963] have shown by careful study of single crystals of α-Fe_2O_3 in the neighbourhood of the Morin temperature at $-13°C$, that there is not a unique angle for all spins which slowly alters with temperature, but that the proportion of spins in the two directions parallel and perpendicular to the [111] plane alters with temperature. Below the transition, Mössbauer spectroscopy has given a spectacular direct demonstration of the spin-flop phenomenon as a function of applied magnetic field, Blum et al. [1965]: the direction of the sublattice magnetization flops from being along the trigonal axis to being perpendicular to this axis when the externally applied field exceeds a critical strength of 67 kOe.

2.4. Line-width and line-shape

It has already been indicated in Section 2.2 that the presence of some Fe^{3+} in nonstoichiometric $Fe_{1-x}O$ distorts the symmetrical quadrupole doublet by broadening and increasing the intensity of one of the components as the result of the presence of an unresolved additional component to the resonance. A more drastic example is given by the work of Banerjee et al. [1967] on disordered iron-titanium spinels such as $Fe_{2.33}Ti_{0.67}O_4$. Here, both ferrous and ferric ions are present on the tetrahedral sites and the same ions, together with the titanium are distributed on the octahedral sites:

$$Fe^{2+}_{0.50}Fe^{3+}_{0.50}[Fe^{2+}_{1.17}Fe^{3+}_{0.17}Ti^{4+}_{0.67}]O_4 .$$

Fig. 7.5 shows that, in the paramagnetic region the line is a broad doublet due to a range of quadrupole interactions, though the presence of Fe^{3+} is already discernable in enhancing the intensity of the low-velocity peak. However, below the Néel temperature there is not only a range of quadrupole interactions, but also a range of hyperfine magnetic fields; the resonance becomes smeared out over a range of some 16–17 mm/sec and is virtually lost in the background noise due to counting statistics. It is effects such as these which severely limit the application of Mössbauer spectroscopy to inverse spinel systems.

2.5. Line intensity

The intensity of a Mössbauer resonance depends on the total number or concentration of the relevant isotope giving rise to the resonance and on its recoil-free fraction. In addition, in single crystals, the intensity may depend

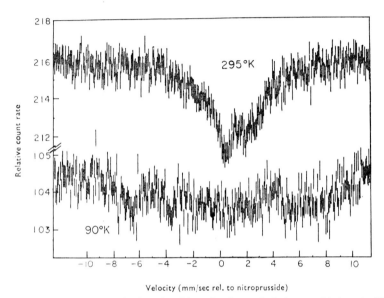

Velocity (mm/sec rel. to nitroprusside)

Fig. 7.5. Mössbauer spectra of a disordered iron titanium spinel above and below the Néel temperature.

on the angle of incidence of the γ-ray since the γ-photons are polarized and the quadrupole and hyperfine Zeeman components have an angular dependence. This latter effect is particularly useful in assigning the various peaks in a complex spectrum and in deducing the direction of the principal component of the electric field gradient tensor.

A recent example of the use of line intensities to determine the relative concentration of ferrous and ferric ions in a nonstoichiometric compound comes from the work of Gibb et al. [1968] on the blue iron titanium double sulphate phase which approximates in composition to $FeTi(SO_4)_3$. Analytical and Mössbauer data indicate substantial variations from the formal cation charges appropriate to an $Fe^{2+}Ti^{4+}$ system and, even when the composition approaches the stoichiometric limit of $FeTi(SO_4)_3$, there are appreciable

References p. 267

concentrations of Fe^{3+}. The phase can be represented as

$$[Fe^{2+}_{1-(x+y)} Fe^{3+}_{(x+y)}][Fe^{3+}_x Ti^{3+}_y Ti^{4+}_{1-(x+y)}][SO_4^{2-}]_3$$

which, in the limit of $1:1$ stoichiometry of iron and titanium becomes

$$[Fe^{2+}_{1-y} Fe^{3+}_y][Ti^{3+}_y Ti^{4+}_{1-y}][SO_4^{2-}]_3.$$

The difficulty of distinguishing analytically between $(Fe^{2+}+Ti^{4+})$ on the one hand and $(Fe^{3+}+Ti^{3+})$ on the other precludes the possibility of assigning oxidation states on the basis of analytical data alone.

When the various Mössbauer lines arise from ions in differing oxidation states, as in the preceding example, no real problems arise because the high-spin ferrous and ferric resonances occur at sufficiently different Doppler velocities to allow adequate resolution and relative intensity measurements. However, if one wishes to determine the site occupancy of a given species, e.g. Fe^{3+}, on tetrahedral and octahedral sites, then the slight differences in chemical isomer shifts in the two situations do not allow complete resolution and one has to be rather more subtle. An ingeneous solution to this problem was recently devised by Armstrong et al. [1966] for the case of γ-Fe_2O_3 which can be represented approximately as a defect spinel $Fe_{2.67} \square_{0.33} O_4$. The compound is ferrimagnetic and a six-line pattern is observed in the Mössbauer spectrum. Suppose, now, the spectrum is recorded in the presence of an externally applied magnetic field: as the tetrahedral A sites and octahedral B sites are magnetically ordered antiparallel to each other, the applied field diminishes the hyperfine field felt by one set of sites and enhances that felt by the other. A field of 17 kOe was applied, giving an added separation of the fields at the tetrahedral and octahedral sites of 34 kOe which was sufficient to resolve them. The relative peak heights of the two six-line spectra then showed that all the A sites were occupied and that all the vacancies were distributed on the B sites, as had previously been predicted. At zero applied field the parameters were then calculated to be

$$\delta_A \ 0.54 \text{ mm/sec}; \quad H_A \ 488 \text{ kOe}$$
$$\delta_B \ 0.68 \text{ mm/sec}; \quad H_B \ 499 \text{ kOe}.$$

The smaller chemical isomer shift and hyperfine magnetic field at the A sites is consistent with the expected increase in covalent character of Fe^{3+} in a tetrahedral environment.

Line intensity also depends on the recoil-free fraction, f, of resonance events at a given site and this, in turn, is related to the lattice dynamics of the crystal and to the mean-square displacement of the Mössbauer atom during the lifetime of the excited state (~ 100 nsec for ^{57}Fe). The full expression for the recoil-free fraction is

$$f = \exp\left\{- E_\gamma^2 \langle x^2 \rangle /(\hbar c)^2\right\}.$$

We may therefore expect lattice defects to profoundly alter the detailed dynamics of the lattice in regions immediately adjacent to a vacancy, interstitial, or altervalent ion and this will influence the integrated intensity of the Mössbauer resonance. A recent example of the influence of localized lattice modes on the recoil-free fraction, f, comes from the work of Ritter et al. [1967]. Instead of using ^{57}Co to generate the ^{57}Fe excited state, they used Coulomb excitation of ground-state ^{57}Fe by 3 MeV α-particles, i.e. the close approach of a positively charged α-particle to the positively charged ^{57}Fe nucleus displaces the atom from a normal lattice site and excites the nucleus to the 14.41 keV Mössbauer level. When this was done with a foil of α-iron, the usual six-line spectrum of ferromagnetic iron was obtained with normal values for both H and f; with an Fe_2O_3 target the intensity dropped by about 50% but the spectrum was otherwise normal. This means that in the metallic phase the displaced iron atoms come to rest on a normal lattice site by replacement collision within a time of ~ 100 nsec whereas in Fe_2O_3 an appreciable fraction of the displaced iron atoms come to rest on abnormal lattice sites where their weaker bonding decreases their recoil-free fraction and they make only a slight contribution to the resonance absorption.

A related experiment has been carried out by Sprouse et al. [1967] who studied the recoil implantation through a vacuum of excited state ^{57}Fe generated by bombardment with 36 MeV oxygen ions. When the Mössbauer nucleus was caught on metallic foils of copper, aluminium, gold, or iron the spectra showed that the implanted atoms were in normal impurity centres in the foils within 100 nsec, whereas recoil implantation into the silicate mineral olivine gave no detectable resonance under the same conditions.

The concluding sections of the chapter apply the general ideas outlined in Sections 1 and 2, to a range of specific examples in order to illustrate further the scope of the results which have been obtained on nonstoichiometric phases by Mössbauer spectroscopy.

References p. 267

3. Illustrative review of results

3.1. Spinels and other oxide systems

The basic diagnostic use of Mössbauer spectroscopy to determine oxidation states and site occupancies in spinels is summarized in Tables 7.1 and 7.2. Taking ferrous spinels first, normal 2,3 spinels such as $FeCr_2O_4$ with Fe^{2+} in the A sites only give a single-line spectrum because the tetrahedral site has cubic symmetry (e.g. Tanaka et al. [1966]. Note that $FeCr_2O_4$ becomes tetragonal below 135°K and the A site loses its cubic symmetry giving a quadrupole split spectrum; below 69°K the compound becomes ferrimagnetic with $H \sim 210$ kOe, Imbert and Martel [1965]). By contrast normal 4,2 spinels such as $GeFe_2O_4$ with Fe^{2+} in the B sites give a two-line spectrum with a large quadrupole splitting because of the large trigonal field at the octahedral site (Rossiter [1966], Mathur et al. [1965]). (The compound becomes antiferromagnetic below 11°K and gives a magnetic hyperfine spectrum with $H = 155$ kOe.) When a 2,3 spinel is partly inverse, for example $FeGa_2O_4$, then two pairs of lines are seen since, in addition to the large quadrupole splitting due to the trigonal field at the B sites, the A sites are no longer cubic due to charge variations on the B sites and a small quadrupole splitting is observed (Rossiter [1966]).

With ferric spinels (Table 7.2) it is not possible to get a normal spinel structure with Fe^{3+} on the A sites because the cationic charge is an odd number, but a disordered spinel such as $Fe^{3+}[Ni^{2+}Al^{3+}]O_4$ has the tetrahedral sites occupied by Fe^{3+}. The disorder on the B sites generates a small e.f.g. and a small quadrupole splitting is observed (Mitzoguchi and Tanaka [1963]). Normal 2,3 spinels such as $ZnFe_2O_4$ with Fe^{3+} on the B sites show a similar spectrum because the e.f.g. from the trigonally distorted octahedral site acting on the spherical d^5 ion, gives a smaller quadrupole splitting than that observed with the d^6 ferrous ion on the B site (Mitzoguchi and Tanaka [1963]). Inverse 2,3 spinels such as $CuFe_2O_4$ have Fe^{3+} on both A and B sites, leading to two superimposed six-line spectra in which only the B site shows a quadrupole effect and the A site has the smaller chemical shift and hyperfine magnetic field as expected (Evans et al. [1966]). Further data can be found in the review by Danon [1968].

These examples lead on to the classic case of magnetite, Fe_3O_4, which can be represented as an inverse spinel phase in which both ferrous and ferric ions are distributed on the octahedral B sites, $Fe^{3+}[Fe^{2+}Fe^{3+}]O_4$. The tetra-

Table 7.1

Mössbauer spectra of ferrous spinels

Spinel type	Site occup.	Example	Spectrum (δ in mm/sec rel. to nitroprusside)	Remarks
normal 2,3	A	$Fe^{2+}[Cr_2^{3+}]O_4$	Single line $\Delta = 0.0$, $\delta = 1.09$	tetrahedral site has cubic symmetry; phase becomes tetragonal below 135°K ($\Delta \neq 0$) and ferrimagnetic below 69°K.
normal 4,2	B	$Ge^{4+}[Fe_2^{2+}]O_4$	Two lines $\Delta = 2.80$, $\delta = 1.38$	large e.f.g. from trigonal distortion on d^6; phase becomes antiferromagnetic below 11°K.
partly inverse 2,3	A and B	$Fe_{1-x}^{2+}Ga_x^{3+}[Fe_x^{2+}Ga_{2-x}^{3+}]O_4$	Two pairs of lines $\Delta_A = 1.2$, $\delta_A = 1.09$; $\Delta_B = 2.9$, $\delta_B = 1.2$	non-cubic A site due to charge variations on B site; trigonal distortion at B site. Note A site has the smaller Δ and lower δ.

Table 7.2

Mössbauer spectra of ferric spinels

Spinel type	Site occup.	Example	Spectrum (δ in mm/sec rel. to nitroprusside)	Remarks
disordered 2,3	A	$Fe^{3+}[Ni^{2+}Al^{3+}]O_4$	Two lines $\Delta = 0.53$, $\delta = 0.44$	cation disorder on B sites removes cubic symmetry of A sites.
normal 2,3	B	$Zn^{2+}[Fe_2^{3+}]O_4$	Two lines $\Delta = 0.36$, $\delta = 0.63$	e.f.g. from trigonally distorted B site acting on spherical d^5.
inverse 2,3	A and B	$Fe^{3+}[Cu^{2+}Fe^{3+}]O_4$ (tetragonal)	Two superimp. six-line spectra A: $\delta = 0.59$, $H = 478$, $\varepsilon = 0.00$ B: $\delta = 0.76$, $H = 514$, $\varepsilon = 0.14$	only B site shows quadrupole effect ($\varepsilon = 0$ for A site may depend on alignment of e.f.g. and H); A site has the smaller H and lower δ.

hedral Fe^{3+} gives one six-line spectrum whereas the "ferrous and ferric" ions randomized on the octahedral sites give an averaged spectrum because, within the lifetime of the excited iron nucleus, any given octahedral ion will have changed between the two oxidation states many times. Below the transition temperature at about 120°K the unit cell becomes orthorhombic, the electron hopping is frozen and the electrical and magnetic properties alter drastically. Simultaneously, the Mössbauer spectrum changes to two new six-line patterns with parameters characteristic of octahedral ferrous on the one hand and ferric ions on the other, the difference between the octahedral and tetrahedral ferric being unresolvable, as already discussed in Section 2.5 for γ-Fe_2O_3. Several groups have investigated this verification at the atomic level of the early ideas of Verwey and his co-workers e.g. Ito et al. [1963]. Typical experimental data are in Table 7.3.

Table 7.3

Mössbauer parameters for Fe_3O_4 (Ito et al. [1963])

Resonant nuclei	δ mm/sec*	ε mm/sec	H kOe	$T°$K
Fe^{3+} on A sites	0.64	+0.02	495⎫	
Fe^{2+}, Fe^{3+} on B sites	1.05	−0.02	470⎭	300°
Fe^{3+} on A and B sites	0.78	+0.05	510⎫	
Fe^{2+} on B sites	1.06	+0.49	470⎭	85°

* relative to hydrated sodium nitroprusside

The site occupancy in $MnFe_2O_4$ has been redetermined by Sawatzky et al. [1967] using the technique already outlined in Section 2.5: an external field of 17 kOe was applied to separate the fields at the A and B sites by a further 34 kOe. The improved resolution gave the ratio of the A:B site occupancies by iron as 0.12 in agreement with the neutron diffraction data. The A and B sites have internal fields of 483 and 430 kOe at room temperature and there is some suggestion of structure in the B spectrum. In a related combined n.m.r. and Mössbauer study by Yasuoka et al. [1967], two different forms of $MnFe_2O_4$ were established: these were formulated as $Mn_{0.8}^{2+} Fe_{0.2}^{3+}$ $[Mn_{0.2}^{3+} Fe_{1.6}^{3+} Fe_{0.2}^{2+}]O_4$ and $Mn_{0.48}^{2+} Fe_{0.52}^{3+} [Mn_{0.06}^{2+} Mn_{0.46}^{3+} Fe_{1.02}^{3+} Fe_{0.46}^{2+}]O_4$. Annealing the latter at 450°C for 5 hours gave the former by cation diffusion and mutual redox interaction between Mn^{3+} and Fe^{2+}.

More complex phases have also been studied. For example, Banerjee et al. [1967] studied the titanoferrite solid solutions $[(1-x)Fe_3O_4 + xFe_2TiO_4]$ by Mössbauer spectroscopy and a variety of other techniques and were able to show that ferric ions had a limited, though not overriding preference for tetrahedral sites. Thus, if there were no site preference there would be a linear variation in the concentration of Fe^{3+} on both tetrahedral and octahedral sites as the end member $Fe^{3+}[Fe^{3+}Fe^{2+}]O_4$ was progressively substituted by $Fe^{2+}[Fe^{2+}Ti^{4+}]O_4$ i.e. the phase would be represented by

$$Fe^{3+}_{1-x}Fe^{2+}_x [Fe^{3+}_{1-x}Fe^{2+} Ti^{4+}_x]O_4 .$$

This was not observed. Nor was there a complete preference of Fe^{3+} for octahedral sites over the whole concentration range as required by the formulations

$$x < 0.5: \quad Fe^{3+}[Fe^{3+}_{1-2x}Fe^{2+}_{1+x}Ti^{4+}_x]O_4$$

$$x > 0.5: \quad Fe^{3+}_{2-2x}Fe^{2+}_{2x-1}[Fe^{2+}_{2-x}Ti^{4+}_x]O_4 .$$

Instead, there was a preference for octahedral sites up to $x = 0.2$ and then indiscriminate substitution up to $x = 0.8$, above which all the Fe^{3+} is located in tetrahedral sites:

$$0 \leqslant x \leqslant 0.2 \quad Fe^{3+}[Fe^{3+}_{1-2x}Fe^{2+}_{1+x}Ti^{4+}_x]O_4$$

$$0.2 \leqslant x \leqslant 0.8 \quad Fe^{3+}_{1.2-x}Fe^{2+}_{x-0.2}[Fe^{3+}_{0.8-x}Fe^{2+}_{1.2}Ti^{4+}_x]O_4$$

$$0.8 \leqslant x \leqslant 1 \quad Fe^{3+}_{2-2x}Fe^{2+}_{2x-1}[Fe^{2+}_{2-x}Ti^{4+}_x]O_4 .$$

As noted in Section 2.4 and Fig. 7.5 the large range of site symmetries and hyperfine fields occasioned by the heterogeneous occupancy of the tetrahedral and octahedral sites gives ultra-broad Mössbauer spectra which severely limit the detailed applicability of the technique in these cases.

The Mössbauer spectra of the ferromagnetic, piezoelectric o-ferrite phases $Fe_xGa_{2-x}O_3$ show two different quadrupole splittings, one of 0.51 mm/sec and the other of 1.08 mm/sec, but both with the same value of 0.70 mm/sec for the chemical isomer shift (Trooster [1965]). The relative intensities of the two spectra are 5:2, which has been found to be consistent with a random distribution of Fe^{3+} and Ga^{3+} ions over sites of octahedral sym-

References p. 267

metry. The unusually large ferric quadrupole splitting observed with the less intense doublet may perhaps be related to the piezoelectric behaviour of this compound.

The preceding examples indicate that Mössbauer spectroscopy can be used not only to study oxidation states, site symmetries, magnetic hyperfine interactions and spin-flop transitions but also that it can be used, in principle, to study rate processes on a time scale of 100 nsec. The fact that a Mössbauer spectrum also gives information about relative concentrations of various species, means that, in addition, it can be used to follow the course of solid-state reactions. Two examples will suffice: the reaction of ZnO and Fe_2O_3 to form the spinel phase $ZnFe_2O_4$, and the disproportionation of the wuestite phase $Fe_{1-x}O$ to give metallic iron and Fe_3O_4. These will now be discussed.

In the first reaction the Mössbauer spectra confirmed that no other phases besides the reactants and the product were formed during the reaction which could therefore be represented as

$$ZnO + Fe_2O_3 = ZnFe_2O_4.$$

The formation of $ZnFe_2O_4$ was followed by studying the Mössbauer spectra (and the X-ray powder diagrams) of quenched samples heated for varying lengths of time at 800°. The results indicated that the cations are still mobile after the oxygen lattice is established and that zinc ions diffuse into their sites more rapidly than do the ferric ions. In short, the rate determining step is the diffusion of iron cations through the lattice.

The wuestite phase is thermodynamically stable only above 570°C; at temperatures a little below this it tends to disproportionate according to the reaction

$$4Fe_{1-x}O = (1-4x)Fe + Fe_3O_4 .$$

$Fe_{1-x}O$ gives an asymmetric doublet spectrum whereas iron gives a six-line spectrum with $H \sim 330$ kOe and magnetite has two overlapping six-line spectra with $H \sim 470$ and 495 kOe. Shechter et al. [1967] have used these variations to follow the reaction by quenching samples which had been held for varying times at a range of temperatures. In principle it should also be possible to study the rate of growth of iron and Fe_3O_4 crystallites since, below a certain size, the volume magnetic exchange energy will be insufficient to effect an ordering of the spins and a single superparamagnetic line would be seen for Fe and one (or two partly resolved) superparamagnetic lines for Fe_3O_4. As the crystallites grow, magnetic ordering would occur and the various six-line spectra would appear.

A related use of Mössbauer spectroscopy is illustrated by the recent work of Bhide et al. [1967] who studied $SrTiO_3$ doped with up to 3% of ^{57}Fe. When the sample had been fired at high temperatures, the room-temperature spectrum was a single quadrupole doublet ($\Delta = 0.9$ mm/sec) centred at a chemical isomer shift value typical of Fe^{3+}. However, after firing at 1000° in an atmosphere of hydrogen, a hyperfine spectrum with $H = 330$ kOe was observed at room temperature (identifying the species as metallic iron which had evidently clustered) plus a central line due to superparamagnetism of small regions of iron dispersed in the crystal matrix. At lower concentration of doped iron, the single line intensified relative to the hyperfine lines indicating a greater proportion of smaller colloidal particles of iron ($\leqslant \sim 10$ nm). Refiring in air regenerated the Fe^{3+} spectrum.

3.2. Sulphides, selenides and tellurides

The Mössbauer systematics derived from the study of oxide systems such as those discussed in the preceding section can be summarized as four rules:

(i) the sequence of chemical isomer shifts is δ (tetrahedral Fe^{3+}) $< \delta$ (octahedral Fe^{3+}) $< \delta$ (tetrahedral Fe^{2+}) $< \delta$ (octahedral Fe^{2+});

(ii) the quadrupole splitting for both Fe^{3+} and Fe^{2+} in sites of accurately cubic symmetry is $\Delta = 0$;

(iii) for non-cubic sites $\Delta(Fe^{3+}) \ll \Delta(Fe^{2+})$;

(iv) the magnetic hyperfine field $H(Fe^{2+}) < H(Fe^{3+})$.

Greenwood and Whitfield [1968] have recently shown that the same trends holds for sulphide phases and that in general the chemical isomer shifts and hyperfine magnetic fields observed in sulphides are less than those in the corresponding oxides because of covalency effects. E.g., $\delta(Fe^{3+}) \sim 0.4$ mm/sec and $\delta(Fe^{2+}) \sim 1.0$ mm/sec in sulphides, compared with values of ~ 0.7 mm/sec and ~ 1.6 mm/sec in oxides. Likewise δ(tetrahedral Fe^{2+}) $< \delta$(octahedral Fe^{2+}) as exemplified by the chemical isomer shifts in $Fe^{2+}[Cr_2^{3+}]S_4$ ($\delta = 0.86$ mm/sec) and $In^{3+}[Fe^{2+}In^{3+}]S_4$ ($\delta = 1.14$ mm/sec). Rule (ii) is illustrated by the zero quadrupole splitting in $Fe[Cr_2]S_4$ and Rule (iii) by the considerable disparity in quadrupole splittings in $K^+Fe^{3+}S_2$ ($\Delta = 0.52$ mm/sec) and in $In^{3+}[Fe^{2+}In^{3+}]S_4$ ($\Delta = 3.27$ mm/sec). Finally, the hyperfine mag-

References p. 267

netic field at Fe^{2+} in sulphides is less than that at Fe^{3+} (e.g. H is 206 kOe in $FeCr_2S_4$ and 360 kOe in $CuFeS_2$). These systematics enabled Greenwood and Whitfield [1968] to make assignments for the formal oxidation states of iron in several minerals where these had been in doubt. Thus, chalcopyrite is $Cu^+Fe^{3+}S_2$ rather than $Cu^{2+}Fe^{2+}S_2$, cubanite is $Cu^+(Fe^{2+}Fe^{3+})S_3$ rather than $Cu^{2+}Fe_2^{2+}S_3$, and stannite is $Cu_2^+Fe^{2+}Sn^{4+}S_4$ rather than $Cu^{2+}Cu^+Fe^{3+}Sn^{2+}S_4$. In this latter case, ^{119}Sn Mössbauer systematics also ruled out the possibility of $Cu_2^{2+}Fe^{2+}Sn^{2+}S_4$.

Little point would be served in reviewing the Mössbauer spectra of sulphide phases in full; many groups have studied them, some in depth, and the results are not dissimilar in principle from those obtained on oxides. For example, the pyrrhotite phase Fe_7S_8 has three distinct iron environments and each shows its own hyperfine magnetic field at 307, 255 and 225 kOe respectively. There is also a spin-flop, Morin transition at 120–130°K which can be studied in detail. It is interesting, however, that although the phase contains over 28% of the iron in the ferric state, as implied by the formula $Fe_2^{3+}Fe_5^{2+} \Box S_8$, there are no separate groups of lines centred on the ferric chemical isomer shift. Presumably a hopping mechanism coalesces some of the Fe^{2+} sites with the Fe^{3+}. A more recent study of Fe_7S_8 by Levinson and Treves [1968] using an externally applied magnetic field confirms both the three distinct iron sites and the absence of specifically ferric resonances, and suggests an allocation of the hyperfine patterns to the various lattice sites.

Studies on $Fe_{1-x}S$ ($0 \leqslant x \leqslant 0.07$) and FeS_2 have also been reported: Ono et al. [1962], Salomon [1960], Imbert et al. [1963] and Kerler et al. [1963]. Of more interest is the recent valuable Mössbauer study of the series of solid solutions formed by the sulphide spinels $FeCr_2S_4$ and $CuCr_2S_4$. Anomalous low Curie constants and magnetic moments, as well as a double reversal of sign of the thermoelectric power have been observed within the $Fe_{1-x}Cu_xCr_2S_4$ system. Various explanations had been profered but the work of Haacke and Nozik [1968] indicate that both divalent and trivalent iron ions exist in the range $0 < x < 0.5$ whereas only trivalent iron exists for $0.5 \leqslant x \leqslant 1$. At the iron-rich end of the phase $Fe_{1-2x}^{2+}Fe_x^{3+}Cu_x^+Cr_2^{3+}S_4^{2-}$ there is a fast electron exchange between Fe^{2+} and Fe^{3+} above the Curie temperature and both iron species are appreciably covalent. Below the Curie temperature nine lines are apparent, suggesting the superposition of two magnetic hyperfine spectra arising from the now distinct ferrous and ferric ions. At the iron-poor end of the system all the iron is in the Fe^{3+} state and the presence of Cu^+ is compensated for by Cr^{4+} according to the formulation $Fe^{3+}_{1-x}Cu^+_x Cr^{3+}_{3-2x} Cr^{4+}_{2x-1} S^{2-}_4$.

Less work has been done on selenides and tellurides, but results have been reported on FeSe, $Fe_{0.9}Se$ (Fatseas [1967]), Fe_9Te_8, Fe_3Te_4 and γ-Fe_2Te_3 (Chappert and Fatseas [1966]). For example, the selenides $Fe_{1-x}Se$ ($0 \leqslant x \leqslant 0.165$) are more complex than the oxides and sulphides. At least two overlapping hyperfine magnetic patterns are seen and the spectra are somewhat dependent on sample preparation. Chemical isomer shifts are lower than in corresponding oxide and sulphide phases because of the increased covalent character. The Mössbauer spectrum of γ-Fe_2Te_3 has shown that, surprisingly, there is no magnetic ordering even down to temperatures as low as 1.5°K.

The limited amount of work published on the Mössbauer spectra of nonstoichiometric phases containing selenium and tellurium stems in part from the increasing practical difficulties of the technique as one goes from oxide systems to sulphides, selenides and finally tellurides. The main problems are as follows:

(i) The increasing mass absorption of selenium and tellurium for soft γ-rays attenuates the beam and reduces the resonant dip. Thus, in oxides, typical resonances are of the order of 10% for natural iron; in sulphides it is $\sim 5\%$, but in selenides it is nearer 2% and in tellurides even less. This increases counting times in order to improve counting statistics. (Random variations in counting are $\sim 100 \sqrt{N}/N$, hence 10^6 counts per channel give a background "noise" of 0.1% which limits the precision of data acquisition from weak resonances.) These difficulties can be overcome to some extent by working at lower temperatures (e.g. liquid helium) or by using enriched absorbers (natural iron has 2.19% ^{57}Fe) but this is expensive and cumbersome.

(ii) A second problem is that, unless the vacancies or interstitials are completely ordered and the phases homogeneous, line-broadening occurs as a result of variations in local electric and magnetic hyperfine fields thus further limiting the detectability of the weak resonances.

(iii) The presence of several differing magnetic sites (e.g. three in Fe_7Se_8) produces complex overlapping patterns of six-line spectra which become increasingly difficult to resolve as the heightened covalency effects diminish the differences in chemical isomer shift of the various sites and also diminish the separation of individual components because of the smaller magnitude of the hyperfine magnetic fields.

(iv) Finally, there is a lack of suitable calibrating compounds amongst the stoichiometric selenium and tellurium compounds of iron which limits the reliability of the Mössbauer systematics. A related difficulty is that, as indicated under (iii), the increasing polarisability of the anions and the trend towards increasing covalency or metallic properties, reduces the diagnostic differences between the so-called ferrous and ferric sites. In short the spacing between the lines diminishes and the lines simultaneously become broader and less intense.

3.3. Alloys and intermetallic phases

By contrast with the sparcity of work on the heavier chalcogenides, the nonstoichiometric alloys of iron and other Mössbauer nuclides with the lighter elements such as boron, carbon, silicon, nitrogen and phosphorus have provided a particularly fruitful area for investigation. Intermetallic phases have also been extensively investigated.

A few typical examples will be selected to indicate the range of studies and the type of information obtainable. First to be considered are the ferromagnetic iron borides in which boron atoms fill the interstices in an otherwise close-packed iron lattice. The first spectra were reported independently by Cooper et al. [1964] and Shinjo et al. [1964], and values for the magnetic fields of Fe_2B and FeB at $300°K$ were found to be 242 kOe and 118 kOe respectively. These values, together with those of the chemical isomer shifts (0.51 and 0.63 mm/sec respectively) when compared with the values for pure iron ($H = 330$ kOe, $\delta = 0.35$ mm/sec) indicate a transfer of electrons from the boron into the metal 3d band. The reduction in the number of unpaired 3d electrons causes the reduction in the hyperfine field, whilst the increased shielding of the s electrons raises the chemical isomer shift. The effects were greater in FeB than Fe_2B because of the higher boron content, and the actual configurations were estimated to be $(2sp)^{\sim 1}(3d)^{8 \cdot 9}(4s)^{\sim 1 \cdot 1}$ and $(2sp)^{\sim 0 \cdot 3}$ $(3d)^{8 \cdot 2}(4s)^{\sim 1}$. The (2sp) band is localised on the boron atoms. This interpretation in terms of electron donation from the interstitial atom to the iron runs counter to naive expectations based on electronegativity considerations (which are here inapplicable) but is paralleled by most of the other alloy systems described.

Mixed borides have been briefly investigated by Cadeville et al. [1965]: Fe_2B and $Fe_{1.2}Co_{0.8}B$ gave fields of 241 kOe and 232 kOe at $300°K$. The iron atom is apparently insensitive to cobalt substitution, although the magnetic susceptibility data suggest that the cobalt is affected by the iron content. Doping of Fe_2B with 2% manganese or 10% manganese causes very little line broadening or satellite formation (Bernas and Campbell [1967]), in contrast to observations on some other types of alloys; it would appear that the internal field arises solely from the d-electron core polarisation term.

Cementite, Fe_3C, is also an interstitial compound, and is ferromagnetic below 210°C. Its spectrum was briefly reported by Shinjo et al. [1964], an internal field of 210 kOe indicating a $3d^8$ configuration for the iron. Ron et al. [1966] found Fe_3C in a cast iron absorber. The field was 208 kOe and $\delta = 0.54$ mm/sec. "Pure" cementite also showed a very small paramagnetic contribution which seemed to be a different form of Fe_3C rather than a superparamagnetic effect.

The $Fe_3(C_{1-x}B_x)$ phase has been investigated by Bernas et al. [1967] who find that the average hyperfine field is proportional to the average moment for each composition. Boron substitution in Fe_3C causes an increase in H_{eff} at room temperature and also a broadening of the lines, with some structure apparent at high boron content. This shows the presence of some short range effects of the interstitial atoms. Chemical isomer shift variations are small and could not be correlated easily with the field changes. Fe_5C_2 has three different iron sites, and fields of 222 kOe, 184 kOe and 110 kOe were allotted to these.

Martensite is a ferromagnetic solid solution of carbon in iron. Gemin and Flinn [1966] studied the effects of ageing at room temperature on an 8.2 atom % solution of carbon in iron. Alterations in the spectrum were attributed to clustering of the carbon atoms so that the initially broadened magnetic spectrum narrowed to that of iron, but with weak satellites to the main peaks. Zemcik [1967] found that, as the carbon content increased from 0 to 7.5 atom % C, there was a gradual increase in the ^{57}Fe hyperfine field from that in pure iron and various mechanisms for this were proposed. Ino et al. [1967] studied a sample of martensite containing 4.2 atom % C in iron: 74% of the absorption intensity was due to iron atoms virtually unaffected by the carbon and the rest was resolved into two hyperfine patterns attributable to irons with first and second coordination neighbours being carbon. A fourth component was paramagnetic austenite. Tempering seemed to cause the loss of the weak hyperfine patterns and the formation of Fe_3C. Similar results were published independently by Gielen and Kaplow [1967] who not only studied the central austenite line in more detail, but also investigated some iron-nitrogen austenite and martensite samples.

Some nitrides based on Fe_4N have been investigated by Shirane et al. [1962]. The Fe_4N structure is based on a face-centred cubic lattice with iron

References p. 267

atoms on all the sites and a nitrogen atom at the body centre. Each corner iron, Fe_A, has 12 Fe neighbours at 2.96 Å and the face-centred iron, Fe_B has 2 nitrogens at 1.90 Å. The room-temperature spectrum comprises two super-imposed hyperfine spectra with fields of 345 kOe and 215 kOe. The intensi-ties (1:3) allow allocation to the A and B sites respectively. The corre-sponding chemical isomer shifts are 0.48 and 0.63 mm/sec giving a net posi-tive shift from iron metal of 0.13 and 0.28 mm/sec. It can be seen that the A site which has only iron atoms in the first coordination sphere has an internal field close to that of iron metal, whereas that of the B site is much smaller. $Fe_{3.6}Ni_{0.4}N$ gives field values and chemical isomer shifts of 363 kOe and 0.48 mm/sec for the A site and 220 kOe and 0.58 mm/sec for the B site. Substitution takes place on the A site with very little effect on the other iron environments. Fe_3NiN has no iron on the A site and shows only one field of 205 kOe with $\delta = 0.63$ mm/sec. Shirane et al. interpret all these data as indicative of electron donation from the nitrogen atoms to the face-centred iron, giving effective electronic configurations of $3d^7 4s^1$ and $3d^8 4s^1$ for the A and B atoms. The Fe_4N measurements were later repeated by Gielen and Kaplow [1967] with similar results.

The iron-silicon phases are intermediate between true interstitial, stoichiometric compounds and metallic alloys. The iron-rich region of the phase is very similar to the aluminium alloys, while there are several stoichio-metric silicon-rich phases which will not be considered here. Samples of the $Fe_{1-x}Co_xSi$ phase with varying values of x all show the same chemical isomer shift. Cobalt substitution either has no effect or there is a chance cancellation. The iron first coordination sphere of 7 silicon nearest neigh-bours in the FeSi lattice is probably partly responsible. Δ, however, decreases from FeSi to CoSi in a non-linear fashion. $Fe_{0.9}Rh_{0.1}Si$ and $Fe_{0.9}Ni_{0.1}Si$ both follow the general trend. The ^{57}Fe atom in nearly pure CoSi does not carry a moment greater than 0.05 μ_B so that the observed susceptibility must arise from the cobalt. $Co_{0.5}Fe_{0.5}Si$ is magnetically ordered below 45°K, but at 4.2°K, H is less than 20 kOe. Various band structures for these compounds are discussed in the original papers.

A major study of the iron-rich alloy phases by Stearns [1963] has already been alluded to on p. 245. The spectra become quite complex as the silicon content increases. The alloys are disordered below 10 atom% Si, and it can be assumed that there is approximately equal probability of any site being occupied by silicon. The observed spectra were still based on the iron metal spectrum but with gross broadening inside the outer lines. This was attributed to a series of hyperfine splittings from iron atoms with 8, 7 or 6 nearest-neigh-bour iron atoms. The internal field is dependent only on the number of iron nearest neighbours and not on the overall silicon content. It appears that the effect of silicon neighbours is not cooperative but independent of each other. The high silicon environments have a more positive chemical isomer shift which is attributed to filling of the 3d-electron orbitals. The ordered alloys from about 14 to 27 atom % Si tend to adopt the Fe_3Si structure which is a body-centred cubic lattice with both iron and silicon on the body-centre sites. Fe_3Si (also briefly studied by Shinjo et al. [1963] and Johnson et al. [1963]) shows two distinct magnetic fields corresponding to the cube-corner (A) and body-centre (B) sites. The larger field of 310 kOe comes from the B site irons with 8 Fe nearest neighbours, while the smaller of 200 kOe is from the A site with 4Fe and 4Si neighbours. Both have isomer shifts more positive than for pure iron. Other stoichiometries gave more complex spectra as a result of different environments at the B site. Fields for 6, 5 and 3 iron neigh-bours were found, and showed the same pattern of behaviour as in the dis-ordered alloys. In the latter there is a decrease in H of $0.08H_{Fe}$ per nearest neighbour Si atom, and $0.14 H_{Fe}$ per Si in the ordered region.

A more detailed analysis of a single crystal of Fe with 4.9% Si has been published by Cranshaw et al. [1966]. Spectra with the crystal magnetised in the [111] and [001] directions showed differences due to anisotropic inter-actions between the impurity atom and its neighbours. Computer data-analysis showed that: (i) it was necessary to assume significant isotropic effects up to the third neighbour shell and an integrated effect from the fourth shell to infinity for both the hyperfine field and the isomer shift; (ii) the effects in the first coordination shell were proportional to $n + \alpha n^2$ (where n is the number of impurity atoms in the shell and $\alpha \sim 0.1$); (iii) strong aniso-tropic effects exist at the second neighbour shell; (iv) the silicon atoms are not randomly distributed but Si–Si nearest neighbour pairs are forbidden.

3.4. Nonstoichiometric hydrides

One final type of Mössbauer experiment which has yet to be mentioned concerns the use of a Mössbauer nuclide like ^{57}Fe to act as a probe to study nonstoichiometric compounds of elements other than iron. A particularly elegant example of this appeared recently in two contrasting papers by Wertheim and Buchanan [1967] on $NiH_{\sim 0.6}$ and by Jech and Abeledo [1967] on $PdH_{\sim 0.7}$. Superficially there is a close parallel between the ranges of the

References p. 267

two nonstoichiometric hydrides and the simplest interpretation is that the electron from the hydrogen goes into the d band of the metal filling the 0.6–0.7 holes which are known to exist. The saturated phases are diamagnetic. Furthermore, neutron diffraction experiments have shown that the hydrogen ions occupy octahedral sites in the face-centred-cubic lattice of both nickel and palladium; that is, approximately a sodium-chloride type structure with 30% vacancies on one of the sublattices. However Mössbauer spectroscopy of suitably doped Ni/H and Pd/H systems show significant differences.

When a nickel foil doped with ^{57}Co is electrolytically charged with hydrogen in a source experiment, the six-line hyperfine spectrum of ^{57}Fe in nickel ($H = 283$ kOe at liquid nitrogen temperature) weakens and a single line due to ^{57}Fe in $NiH_{0.6}$ appears. The most striking feature of the spectra is the appearance of only two distinct spectra in the whole composition range, viz. ^{57}Fe in Ni and ^{57}Fe in $NiH_{0.6}$ as shown in Fig. 7.6 (Wertheim and Buchanan [1967]).

The single line does not split even at 4°K and the hyperfine field was independent of the amount of hydrogen in the sample. It is difficult to escape the conclusion that the hyperfine magnetic field stems from iron in regions containing essentially no hydrogen and that the single line arises from iron in regions containing a constant composition of $NiH_{0.6}$. The variation in composition of the foil is due to the variation in the proportions of these two phases which, in turn is reflected in the relative intensities of the six-line and the single-line spectra. This is in contrast to the behaviour of the Cu/Ni system in which copper, rather than hydrogen supplies the extra electron, Wertheim and Wernick [1961]: here addition of copper progressively collapses the hyperfine field of iron in the host alloy.

The chemical isomer shift of the ^{57}Fe/$NiH_{0.6}$ line relative to that in ^{57}Fe/Ni is also most interesting. The increase in δ from 0.14 to 0.63 mm/sec corresponds to a decrease in s electron density at the ^{57}Fe nucleus and is twice the shift produced by the extreme of the Ni/Cu alloy system. Calculations suggest that iron still has unpaired d electrons in non-magnetic $NiH_{0.6}$ and this was confirmed by applying an external field of 50 kOe at 4.2°. A parallel Mössbauer study of an iron foil subjected to cathodic charging gave a spectrum identical to that of untreated iron, confirming that very little hydrogen remains in solid solution; this accords with the known equilibrium of hydrogen in iron which is $\sim 10^{-5}$H/Fe at 300°C and is even lower at lower temperatures.

The experiments of Jech and Abeledo [1967] on the hydrogenation of

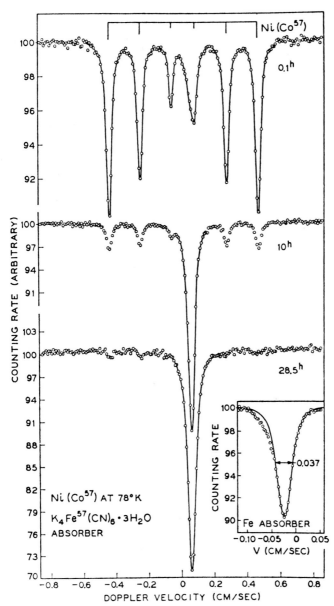

Fig. 7.6. Mössbauer spectrum of ^{57}Fe in $Ni/NiH_{0.6}$.

Fe/Pd alloys gave quite different results. Using alloys with 11% and 15% Fe, only one homogeneous, nonstoichiometric phase was obtained on progressive hydrogenation. The dissolved hydrogen decreased the Curie ferromagnetic temperature but left the observed hyperfine field at the iron atom unchanged; the susceptibility of each Fe/Pd alloy decreases with concentration down to a residual value which depends on the iron concentration in the alloy. This suggests a progressive filling of the 4d band (Pd) by electrons from the hydrogen dissolved in atomic form whilst the magnetic moment at the iron atoms remains constant. In other words, the electrons from hydrogen do not interact significantly with the iron 3d electrons. Despite this, there is a slight increase of about 0.1 mm/sec in the chemical isomer shift of Fe/Pd alloys on saturation with hydrogen; this is ascribed to the known expansion of the lattice on hydrogenation rather than to an increase in iron 3d electrons which seems to be precluded by the magnetic data.

4. Conclusion

This review has outlined the ways in which Mössbauer spectroscopy can be applied to the study of nonstoichiometric phases. The technique has been dominated by the ^{57}Fe resonance in work reported so far but other elements are potentially capable of giving important results and these are currently being investigated.

In favourable cases, the chemical isomer shift, δ, gives information about the charge states of both host ions and defect ions and also whether there is rapid electron exchange between them. The presence of unusual oxidation states such as Fe^+ and Fe^{4+} can also be detected. Subtler changes in δ and in the hyperfine magnetic field H lead to a quantitative determination of the distribution of iron atoms on sites of different symmetry. Diagnostic use can also be made of the quadrupole splitting parameter as exemplified by the critical dependence of the Mössbauer spectrum on the details of site occupancy in normal, inverse, and disordered spinels containing Fe^{2+} and Fe^{3+}.

Temperature-dependence studies can lead to the enthalpy of interaction of defects. Changes in the recoil-free fraction give information about the lattice dynamics of defects and the use of line intensities can also be used to monitor the course of solid-state chemical reactions.

Various sites within a lattice are frequently associated with individual values of the hyperfine magnetic field and this permits the determination of variations of site occupancy with phase concentration. *H* may also change with composition of the phase, giving further information about the electronic structure and exchange interactions. Temperature dependence of hyperfine magnetic fields leads to the determination of Curie and Néel temperatures and their variation with composition. The ability to determine the angle between the hyperfine magnetic field and the principal direction of the electric field gradient tensor makes Mössbauer spectroscopy a valuable technique for studying spin-flop phenomena at the Morin temperature.

The main limitation of Mössbauer spectroscopy as applied to non-stoichiometric phases centres on the limited number of elements which can currently be studied conveniently. In addition, the problem of resolution of numerous overlapping spectra sometimes complicates the analysis and diminishes confidence in the detailed assignments of bands. This problem is exacerbated when large, highly polarisable elements such as selenium and tellurium are present in the phase being studied.

Despite these difficulties it remains true that Mössbauer spectroscopy has added spectacularly to our knowledge of defect solid phases during the past decade. No other single technique gives so much diverse and intimate information from a single experiment and the continual advances in radio-chemical and electronic techniques encourage the belief that the technique will become more readily applicable to a wider range of elements in the future.

The author acknowledges with pleasure the many stimulating and helpful discussions he has had with his colleague Dr. T. C. Gibb.

List of symbols

A	tetrahedral lattice sites in spinel-type phase
B	octahedral lattice sites in spinel-type phase
c	speed of light (in mm/sec)
e	electron charge
E_γ	energy of a γ-transition or γ-photon (in keV)
E_R	free-atom recoil energy (in eV)
f	recoil-free fraction
g_0	nuclear Landé splitting factor for ground state
g_1	nuclear Landé splitting factor for first excited state
h	Planck constant
$\hbar$	$h/(2\pi)$
H	nuclear hyperfine magnetic field (in kOe)

References p. 267

I	nuclear spin quantum number
m	component of nuclear spin momentum in direction of Z-axis
M	mass of an atom or assembly of atoms (in a.m.u.)
N	number of γ-photon counts recorded
Q	nuclear quadrupole moment
q	electric field gradient
T	temperature
$t_{\frac{1}{2}}$	half-life of nuclear excited-state
v	Doppler velocity (in mm/sec)
x	Cartesian coordinate
x	stoichiometry variable e.g. $Fe_{1-x}O$
$\langle x^2 \rangle$	mean square displacement of Mössbauer nuclide during life-time of excited state
y	Cartesian coordinate
y	stoichiometry variable e.g. $[Fe_x^{3+}Ti_y^{3+}Ti_{1-x+y}^{4+}]$
z	Cartesian coordinate, principal crystal axis or direction of electric field gradient
α	phase designation e.g. α-Fe_2O_3
α	alpha particle, $_2^4He$
β_N	nuclear Bohr magneton $e\hbar/(2Mc)$
γ	phase designation e.g. γ-Fe_2O_3
γ	gamma-photon
$(1-\gamma_\infty)$	Sternheimer shielding factor
Γ	line-width at half-height (in mm/sec or eV)
δ	chemical isomer shift (in mm/sec relative to $Na_2[Fe(CN)_5NO]2H_2O$ as zero, for ^{57}Fe)
Δ	electrical quadrupole splitting (in mm/sec)
ε	electric quadrupole perturbation of nuclear hyperfine magnetic spacings
$|\psi_s(0)|^2$	electron density at the resonant nucleus
$^\circ$	degrees Celsius
$^\circ K$	degrees Kelvin
$\square$	vacant cation site
$|\,\rangle$	nuclear state or electronic state identified by the symbols enclosed e.g. $|\pm\frac{1}{2}\rangle$ or $|xy\rangle$

References

ARMSTRONG, R. J., A. H. MORRISH and G. A. SAWATZKY, 1966, Phys. Letters **23**, 414.

BANERJEE, S. K., W. O'REILLY, T. C. GIBB and N. N. GREENWOOD, 1967, J. Phys. Chem. Solids **28**, 1323.

BERNAS, H. and I. A. CAMPBELL, 1967, Phys. Letters **24A**, 74.

BERNAS, H., I. A. CAMPBELL and R. FRUCHART, 1967, J. Phys. Chem. Solids **28**, 17.

BHIDE, V. G., H. O. BHASIN and G. K. SHENOY, 1967, Phys. Letters **24A**, 109.

BLUM, N., A. J. FREEMAN, J. W. SHARRER and L. GRODZINS, 1965, J. Appl. Phys. **36**, 1169.

CADEVILLE, M. C., R. WENDLING, E. FLUCK, P. KUHN and W. NEUWIRTH, 1965, Phys. Letters **19**, 182.

CHAPPERT, J. and G. A. FATSEAS, 1966, Compt. Rend. **262B**, 242.

COOPER, J. D., T. C. GIBB, N. N. GREENWOOD and R. V. PARISH, 1964, Trans. Faraday Soc. **60**, 2097.

CRANSHAW, T. E., C. E. JOHNSON, M. S. RIDOUT and G. A. MURRAY, 1966, Phys. Letters **21**, 481.

DANON, J., 1968, ^{57}Fe: Metal, Alloys, and Inorganic Compounds, in: Chemical Applications of Mössbauer Spectroscopy, eds. V. I. Goldanskii and R. H. Herber (Academic Press, New York) pp. 159–267.

ELIAS, J and J. W. LINNETT, 1969, in the press.

EVANS, B. J., S. HAFNER and G. M. KALVIUS, 1966, Phys. Letters **23**, 24.

FATSEAS, G. A., 1967, Compt. Rend. **265B**, 1073.

GALLAGHER, P. K., J. B. MCCHESNEY and D. N. E. BUCHANAN, 1964, J. Chem. Phys. **41**, 2429.

GEMIN, J. M. and P. A. FLINN, 1966, Phys. Letters **22**, 392.

GIBB, T. C., N. N. GREENWOOD, A. TETLOW and W. TWIST, 1968, J. Chem. Soc. A 2955.

GIELEN, P. M. and R. KAPLOW, 1967, Acta Metallurgica **15**, 49.

GOLDANSKII, V. I. and R. H. HERBER (Eds.), 1968, Chemical Applications of Mössbauer Spectroscopy (Academic Press, New York).

GREENWOOD, N. N., 1970, Mössbauer Spectroscopy, in: Physical Chemistry, An Advanced Treatise, Vol. 4 (Academic Press, New York) Chapter 11.

GREENWOOD, N. N. and H. J. WHITFIELD, 1968, J. Chem. Soc. A 1697.

HAACKE, G. and A. J. NOZIK, 1968, Solid State Comm. **6**, 363.

HANNAFORD, P., C. J. HOWARD and J. W. G. WIGNALL, 1965, Phys. Letters **19**, 257.

IMBERT, P., 1966, Compt. Rend. **263**, B184.

IMBERT, P. and A. GERARD, 1963, Compt. Rend. **257**, 1054.

IMBERT, P., A. GERARD and M. WINTERBERGER, 1963, Compt. Rend. **256**, 4391.

IMBERT, P. and E. MARTEL, 1965, Compt. Rend. **261**, 5404.

INO, H., T. MORIYA, F. E. FUJITA and Y. MAEDA, 1967, J. Phys. Soc. Japan **22**, 346.

ITO, A., K. ONO and Y. ISHIKAWA, 1963, J. Phys. Soc. Japan **18**, 1465.

JECH, A. E. and C. R. ABELEDO, 1967, J. Phys. Chem. Solids **28**, 1371.

JOHNSON, C. E., M. S. RIDOUT and T. E. CRANSHAW, 1963, Proc. Phys. Soc. **81**, 1079.

KIRLER, W., W. NEUWIRTH, E. FLUCK, P. KUHN and B. ZIMMERMANN, 1963, Z. Physik **173**, 321.

LEVINSON, L. M. and D. TREVES, 1968, J. Phys. Chem. Solids **29**, 2227.

MATHUR, H. B., A. P. B. SINHA and C. M. YAGNIK, 1965, Solid State Comm. **3**, 401.

MIZOGUCHI, T. and M. TANAKA, 1963, J. Phys. Soc. Japan **18**, 1301.

MÖSSBAUER, R. L., 1958, Naturwissenschaften **45**, 538; Z. Physik **151**, 124.

MULLEN, J. G., 1963, Phys. Rev. **131**, 1415.

ONO, K., A. ITO and E. HIRAHARA, 1962, J. Phys. Soc. Japan **17**, 1747.

RITTER, E. T., P. W. KEATON, Y. K. LEE, R. R. STEVENS and J. C. WALKER, 1967, Phys. Rev. **154**, 287.

RON, M., H. SHECHTER, A. A. HIRSCH and S. NIEDZWIDZ, 1966, Phys. Letters **20**, 481.

ROSSITER, M. J., 1966, Phys. Letters **21**, 128.

SALOMON, I., 1960, Compt. Rend. **250**, 3828.

SAWATZKY, G. A., F. VAN DER WOUDE and A. H. MORRISH, 1967, Phys. Letters **25A**, 147.

SHECHTER, H., P. HILLMAN and M. RON, 1966, J. Appl. Phys. **37**, 3043.

SHINJO, T., F. ITOH, H. TAKAKI, Y. NAKAMURA and N. SHIKAZONO, 1964, J. Phys. Soc. Japan **19**, 1212.

SHINJO, T., Y. NAKAMURA and N. SHIKAZONO, 1963, J. Phys. Soc. Japan **18**, 797.

SHIRANE, G., D. E. COX and S. L. RUBY, 1962, Phys. Rev. **125**, 1158.

SHIRANE, G., W. J. TAKEI and S. L. RUBY, 1962, Phys. Rev. **126**, 49.

SPROUSE, G. D., G. M. KALVIUS and S. S. HANNA, 1967, Phys. Rev. Letters **18**, 1041.

STEARNS, M. J., 1963, Phys. Rev. **129**, 1136.

TANAKA, M., T. TOKORO and Y. AIYAMA, 1966, J. Phys. Soc. Japan **21**, 262.

TROOSTER, J. M., 1965, Phys. Letters **16**, 21.

WERTHEIM, G. K. and D. N. E. BUCHANAN, 1967, J. Phys. Chem. Solids **28**, 225.

WERTHEIM, G. K. and J. H. WERNICK, 1961, Phys. Rev. **123**, 755.

WERTHEIM, G. K., J. H. WERNICK and D. N. E. BUCHANAN, 1966, J. Appl. Phys. **37**, 3333.

YASUOKA, H., A. HIRAI, T. SHINJO, M. KIYAMA, Y. BANDO and T. TAKADA, 1967, J. Phys. Soc. Japan **22**, 174.

ZEMCIK, T., 1967, Phys. Letters **24A**, 148.

AUTHOR INDEX

FORMULA INDEX

SUBJECT INDEX

278

thermal stability of defect —, 186
Spin flop phenomena, 246, 256, 265
Spinodal point, 12
Spinodal unmixing, 12
  Cahn's theory of — —, 13
Spin ordering in Fe_7S_8, 17
Splitting, electrical quadrupole —, 241
  nuclear Landé — factor, 244
  quadrupole —, 244, 250, 253–255, 264
  Zeeman —, 244
Stable vacancy-free phase, 196
Stability limits of nonstoichiometric
  phases, 11, 12
  thermal — of defect spinels, 186
  thermal — of the defect form of
    $Co_3V_2O_8$, 192, 193
  thermal — of vacancies containing
    compounds, 179–181, 190, 193
  thermal — of the two defect olivine
    phases, 197
Stacking fault, 112, 113
Stannite, 255
Statistical mechanics of crystals, 1
Sternheimer shielding factor, 243
Stirling's formula, 175
Stoichiometric crystal, 10, 41, 44, 132,
  134
— defect, 42, 53
Stoichiometric deviation, 34
— — of KBr, 38
— — of PbS, 39
Stoichiometric phase, 215
— point, 7, 10, 37, 38
  two-dimensional — phases, 216
— variations, 39
Stoichiometry, deviation from —, *see*
  deviation from stoichiometry
— of the adsorbed layer, 212
Strain energy, 172
Strain field, 146, 147
  averaged potential of — —, 147
— — energy, 146
Structure, averaged —, 135, 136, 138, 139,
  143, 145, 147, 167, 168
  B 1 —, 64
  B 8 —, 15, 65, 109

band —, 141
block —, 19, 21, 22, 24
C 6 —, 15, 65, 112
corundum —, 163
defect —, 26, 219, 222, 232, 233, 235,
  236, 238
domain —, 175, 176
double shear —, 19
electronic —, 93, 99, 141, 265
— elements of a crystal, 10
fluorite —, 65, 103, 108
intergrowth —, 22
inverse spinel —, 152, 157
NiAs —, 15, 114
nonstoichiometric — types, 132
normal spinel —, 152, 250
— of spinels, 149, 152, 153, 163, 164, 191
olivine —, 152
ordered —, 122, 133, 138, 145, 153,
  156, 158
pyrochlore —, 65, 94, 106
regular intergrowth —, 64
ReO_3 —, 19
rock salt —, 163, 175
shear —, 18, 114, 122, 126
spinel —, 149, 152, 153, 163, 164, 191
Structures, coherent growth of —, 64
Subdivision equilibrium, 66, 73
— reaction, 69
Sublattice, 27
Submicrodomains, 104
Submicroheterogeneity, 62
Substitution, 77, 93, 134, 136, 139
  nonstoichiometric —, 137
Substructure, 180
Sulphides, Mössbauer spectra of —, 255
Sulphide spinels, 256
Sulphur, adsorption isotherm of —, 210
  adsorption of — on various metals,
    209
  surface concentration of —, 211
  uniform distribution of — atoms, 212
Superparamagnetism, 254, 255
Superstructure, 15, 78, 91, 125, 134, 136,
  138, 144, 153, 154, 164, 174
Surface compound, 214–216